Understanding Criminal Liability for Workplace Harm in Australia

Courts in Australia routinely prosecute workplace harm as criminal conduct. Leaders, officers, and organisations are held accountable not only for what they did, but for what they failed to foresee, prevent, or control. This book examines how courts determine criminal responsibility after workplace harm, tracing the judicial reasoning that separates conviction from acquittal.

The book is structured around real, publicly available court decisions and prosecution outcomes and cases related to workplace safety drawn from jurisdictions across Australia. It covers three categories of prosecution: fatalities and industrial manslaughter cases, where physical harm exposes systemic failure; officer prosecutions, where distance, delegation, and governance are tested against statutory duties; and psychosocial prosecutions, where courts assess culture, workload, and behaviour as safety systems capable of causing harm. Each chapter analyses a specific case, tracing the prosecution's argument, the defence position, the judicial reasoning, and the outcome. The book's central question is one that practitioners regularly face but rarely have answered with precision: why do prosecutions succeed in some cases and fail in others? The answer is that courts apply the concept of reasonably practicable with remarkable consistency, and this book makes those patterns visible and accessible to every person with a duty of care.

Understanding Criminal Liability for Workplace Harm in Australia: The Reasonably Practicable Standard has been written for safety leaders, officers, directors, and regulators as well as lawyers, who make decisions under uncertainty and who may later find those decisions scrutinised in court.

Understanding Criminal Liability for Workplace Harm in Australia

The Reasonably Practicable Standard

Tom Bourne

CRC Press
Taylor & Francis Group
Boca Raton London New York

CRC Press is an imprint of the
Taylor & Francis Group, an **informa** business

First edition published 2027
by CRC Press
2385 NW Executive Center Drive, Suite 320, Boca Raton FL 33431

and by CRC Press
4 Park Square, Milton Park, Abingdon, Oxon, OX14 4RN

CRC Press is an imprint of Taylor & Francis Group, LLC

ISBN: 978-1-041-43093-3 (hbk)
ISBN: 978-1-041-43259-3 (pbk)
ISBN: 978-1-042-03199-3 (ebk)

DOI: 10.1201/9781042031993

Typeset in Times
by KnowledgeWorks Global Ltd.

Contents

About the Author

Tom Bourne is an Australian work health and safety professional with more than two decades of experience advising organisations on risk management, safety culture, and legal accountability. His work spans high-hazard industries including construction, manufacturing, resources, and professional services. He is the host of Voices from the Edge, a podcast exploring the human dimensions of workplace harm through conversations with regulators, legal practitioners, frontline workers, and organisational leaders. His approach to safety is built on first principles: understanding how courts interpret duty, how organisations create the conditions for harm, and how leaders at every level can think more clearly about decisions that will later be scrutinised. He is a recognised voice in the Australian WHS community.

Preface

This book examines how Australian courts decide criminal responsibility after workplace harm. It is written for leaders, officers, directors, regulators, lawyers, and practitioners who must make decisions under uncertainty and may later find those decisions examined in court.

The cases analysed here are drawn from publicly available court decisions and regulator prosecution outcomes across Australian jurisdictions. Where full judicial reasons are available, they are relied upon directly. Where they are not, official prosecution summaries and sentencing remarks are used. No facts have been invented, no motives inferred beyond the evidence, and no conduct embellished for effect.

This book is not an advocacy text. It does not argue for harsher penalties or broader enforcement. Nor does it seek to minimise harm or excuse failure. Its purpose is explanatory. It asks a practical question that leaders regularly face but rarely have answered clearly: Why do prosecutions succeed in some cases and fail in others?

To answer that question, the book is structured in three parts. The first examines fatalities and industrial manslaughter prosecutions, where physical harm exposes systemic weakness. The second examines officer prosecutions, where distance, delegation, and governance are tested against statutory duties. The third examines psychosocial prosecutions, where courts assess culture, workload, and behaviour as safety systems capable of causing harm.

Throughout, the focus remains on judicial reasoning rather than regulatory rhetoric. Courts, not guidance documents, determine liability. Understanding how courts think is therefore essential for anyone charged with responsibility for work health and safety.

This book does not promise certainty. Safety decisions are rarely binary, and courts acknowledge complexity. What it offers instead is clarity: patterns of reasoning, boundaries of liability, and the conditions under which responsibility crystallises into criminal accountability.

If readers finish this book with a clearer understanding of what courts expect, and why, it will have achieved its purpose.

PART ONE

Fatalities and Industrial Manslaughter

When the Risk Was Already Known 1

R v Narellan Pools Pty Ltd (Queensland)

WORK, PLANT, AND PROXIMITY

Narellan Pools Pty Ltd operated a manufacturing and distribution facility at Stapylton in south-east Queensland. The business specialised in the production and transport of large prefabricated swimming pools. These pools were substantial in size and weight, requiring mechanical assistance to move, store, and load them for delivery.

The work environment was typical of many industrial yards. Mobile plant such as cranes and forklifts were used daily. Workers on foot performed tasks around the same spaces: guiding loads, preparing materials, checking clearances, and assisting with staging and dispatch. There was no clean separation between 'plant work' and 'people work'. The two occurred together, as part of ordinary operations.

This point matters more than it might initially appear. Courts do not assess risk in the abstract. They look closely at how work actually occurs. Where the interaction between mobile plant and pedestrians is embedded in routine operations, the expectation of control is higher. The hazard is not occasional or unexpected; it is structural.

In industrial safety law, mobile plant–pedestrian interaction is one of the most well-documented causes of workplace fatalities. Regulators, industry

DOI: 10.1201/9781042031993-2

bodies, and courts treat it as a classic, high-consequence risk. The controls are neither obscure nor controversial. They include traffic management plans, physical separation, exclusion zones, spotters, enforced procedures, and supervision. These measures are widely understood and widely applied.

Against this background, the work at Narellan Pools was not unusual. What would later become critical was not what the business did, but how it controlled the risks that came with it.

THE INCIDENT

On 19 August 2021, Kamndiin Turner, 42, of Ipswich, was fatally injured at the Stapylton facility. The worker was on foot in the vicinity of a mobile crane while it was operating. During its movement, the crane struck the worker, who was then run over. The injuries were catastrophic, and the worker died at the scene.

There was no suggestion that the crane malfunctioned. It did what cranes do: it moved in response to operator input. The fatal outcome did not arise from mechanical failure or unforeseeable defect. It arose from the way people and plant were permitted to occupy the same space.

This distinction shaped everything that followed.

In workplace prosecutions, incidents involving unexpected equipment failure often turn on maintenance regimes and inspection records. Incidents involving functioning equipment turn on systems of work. In this case, the system controlling crane movements in proximity to pedestrians was the focal point.

REGULATORY FOCUS AND EVIDENCE

The investigation quickly concentrated on how crane operations were planned and controlled. This included examination of:

- Whether exclusion zones existed around the crane's operating envelope
- How workers were prevented from entering those zones
- Whether traffic management arrangements were documented and enforced
- The level of supervision present during crane operations
- Whether prior concerns had been raised about crane movements or speed

What emerged was not a single dramatic failure, but an accumulation of absences. Controls that are normally expected in environments where mobile plant operates near people were either missing or ineffective.

Importantly, the risk was not merely systemic. The evidence revealed a specific human factor that management had failed to address. The crane operator had a documented history of driving the crane at speed. Colleagues had warned him on multiple occasions to slow down. Three months before the fatal incident, he struck a parked car with the crane. On the morning Kamndiin Turner died, a colleague again challenged him about his speed. His manner of operation had been raised directly with the facility manager before the incident. None of these warnings produced a meaningful response. This transformed the case from a straightforward system failure into a known-human-factor-unaddressed-by-supervision matter. The facility manager's supervisory inaction was on the record and materially shaped the court's analysis of culpability.

In legal terms, that matters a great deal.

THE CHARGE: INDUSTRIAL MANSLAUGHTER

Narellan Pools Pty Ltd was charged with industrial manslaughter under Section 34C of the *Work Health and Safety Act 2011 (Qld)*. The offence requires proof that a duty holder negligently caused the death of a worker.

Negligence in this context does not require intention to cause harm. It requires a failure that involves such a great falling short of the standard of care that it warrants criminal sanction.

The prosecution also alleged a breach of the primary duty of care under Section 19 of the Act, which requires a person conducting a business or undertaking to ensure, so far as is reasonably practicable, the health and safety of workers.

HOW THE PROSECUTION BUILT THE CASE

The prosecution's case was notable for what it did *not* rely on. It did not depend on expert testimony to establish exotic or technical hazards. It did not argue that the risk was novel or emerging. Instead, it relied on a proposition courts have accepted repeatedly:

Interaction between mobile plant and pedestrians is a well-known, foreseeable cause of workplace death.

From that starting point, the prosecution constructed its argument around foreseeability and control.

It argued that Narellan Pools Pty Ltd:

- Knew, or ought reasonably to have known, that allowing workers to be on foot near operating cranes created a risk of serious injury or death
- Failed to implement effective traffic management systems to separate people from plant
- Did not enforce exclusion zones around crane operations
- Failed to adequately supervise and control crane movements
- Permitted unsafe proximity to persist as part of ordinary work

Crucially, the prosecution framed the incident as systemic, not individual. The focus was not on blaming the crane operator. It was on the environment in which the operator worked. Courts are consistently wary of narratives that reduce fatal incidents to 'operator error'. Where the system permits error to have catastrophic consequences, liability attaches to the system.

The prosecution also emphasised that the controls required to prevent the incident were basic, inexpensive, and widely recognised. This is central to the concept of 'reasonably practicable'. Where controls are obvious and readily available, failure to implement them is difficult to justify (Office of the Work Health and Safety Prosecutor, 2024).

One further detail warrants attention. Prior to the incident, Narellan Pools had engaged qualified external work health and safety consultancy firms to identify risks across its operations. Those consultants reviewed the company's sites but did not identify the crane–pedestrian interaction as requiring urgent action. This did not assist the company's legal position. Courts do not treat the engagement of consultants as a substitute for operational control, nor does a consultant's failure to identify a risk excuse the business from its primary duty. What the consultants' involvement illustrates is that risk identification is a persistent and active obligation, not an event discharged by commissioning a single review.

THE DEFENCE POSITION

Narellan Pools Pty Ltd entered a guilty plea. As a result, there was no contested trial on liability. The defence did not seek to argue that the risk was unforeseeable or that adequate controls existed.

Instead, defence submissions focused on mitigation. These included expressions of remorse, cooperation with the investigation, steps taken after the incident to improve safety, and submissions regarding the company's financial circumstances.

This approach was pragmatic. In cases involving mobile plant and pedestrian fatalities, arguments that the risk was not known or not controllable rarely succeed. Early guilty pleas are often the only rational strategy to reduce penalty exposure.

JUDICIAL REASONING

In sentencing the company, the court accepted the prosecution's characterisation of the offence.

The court emphasised several key points.

First, the risk was obvious. Courts have little patience for arguments that treat well-known hazards as unforeseeable. The interaction of cranes and pedestrians falls squarely within the category of risks that demand active control.

Second, the absence of effective controls was critical. The court distinguished between having informal practices and having enforced systems. Where exclusion zones exist only in theory, they do not reduce risk in practice.

Third, the court rejected any suggestion that the incident was a tragic accident in the sense of being unavoidable. It was, instead, the realisation of a known risk that had not been adequately managed.

In determining penalty, the court explicitly measured the case against *R v Brisbane Auto Recycling Pty Ltd* [2020] QDC 113, where $3 million had been imposed for industrial manslaughter. The sentencing judge found that Narellan Pools' operations were vastly different from Brisbane Auto Recycling, with materially less flagrant conduct. Nonetheless, a substantial fine remained warranted given general deterrence. The $1.5 million figure, exactly half the Brisbane Auto penalty, was a calibrated signal: the gravity of conduct determines penalty, not merely the fact of death. Less flagrant conduct attracts a proportionately lower fine, but not an inconsequential one.

In assessing penalty, the court placed significant weight on general deterrence and denunciation. Industrial manslaughter offences are designed to signal that certain failures are no longer to be treated as regulatory missteps but as serious criminal conduct.

SENTENCE

Narellan Pools Pty Ltd was convicted of industrial manslaughter and fined **$1.5 million**.

The fine reflected:

- The loss of life
- The seriousness of the negligence
- The availability of reasonably practicable controls
- The need to deter similar conduct across the industry

While the guilty plea and cooperation were acknowledged, they did not diminish the objective gravity of the offence.

WHY THIS CASE MATTERS

This case captures the core logic that underpins the entire book.

The prosecution did not succeed because something rare or extraordinary occurred. It succeeded because something ordinary was allowed to remain dangerous.

The risk was known
The controls were basic
The failure was tolerated

Courts assess liability against what was foreseeable at the time, not what became obvious afterward. Hindsight may illuminate a risk, but it does not ground the conviction. What grounds it is the failure to act on what a reasonable duty holder ought to have known and could have controlled. That distinction protects organisations from liability for genuine surprises while holding them accountable for the predictable consequences of known and unmanaged risk.

For leaders, the lesson is clear and uncomfortable. Where risk is embedded in daily work, reliance on informal practice, experience, or good intentions is not enough. Reasonably practicable measures must be implemented, enforced, and verified before harm occurs.

This is the standard by which conduct is judged when the consequences are fatal.

The broader principle remains: reasonably practicable requires action before harm occurs. The controls that would have protected Kamndiin Turner were basic, available, and inexpensive. Their absence was not a resource problem. It was a leadership decision, and courts treated it accordingly.

The Brisbane Auto Recycling comparison demonstrates how courts calibrate penalties. Fines are not arbitrary. They are set by reference to the gravity of conduct, the availability of controls, and comparable cases. A $1.5 million fine calibrated against a $3 million comparator still represents a criminal conviction for the death of a worker. Less flagrant is not the same as acceptable.

The external consultants' engagement adds a dimension many organisations overlook. A third-party safety review is not legal insulation. Consultants can identify hazards, but they cannot substitute for ongoing operational vigilance. Where a review misses a known recurring risk, liability remains with the business. Commissioning a review and then doing nothing further is the appearance of due diligence, not the substance of it.

For leaders, this case carries lessons that extend beyond mobile plant management. The death of Kamndiin Turner illustrates how individual conduct within an organisation can become a systemic failure. The crane operator's dangerous driving pattern was known, raised with management, and tolerated. When a supervisor knows a worker is behaving dangerously and takes no meaningful action, the law treats that inaction as part of the employer's failure, not a separate matter.

REFERENCES

Office of the Work Health and Safety Prosecutor (2024) *R v Narellan Pools Pty Ltd—Industrial manslaughter conviction*. Queensland Government.

Work Health and Safety Act 2011 (Qld), ss 19, 34C.

Negligence Without Excuse

2

R v Owen (Queensland)

A SMALL BUSINESS, FAMILIAR WORK, AND INFORMAL PRACTICE

Jeffrey Owen operated a small business involved in the transport, handling, and movement of heavy equipment, including industrial generators. Like many operations of its kind, the work was practical, repetitive, and grounded in experience rather than formal engineering controls. Forklifts were used routinely to lift and position generators for loading and unloading.

The environment was not chaotic. It was familiar. Workers understood the equipment, the loads, and the tasks. That familiarity, however, is often where risk begins to hide. When work is done the same way day after day, informal practices can gradually replace formal systems. What 'works' becomes what is accepted.

In this case, generators were lifted on forklift tines without being mechanically restrained. This practice was not occasional. It was how the work was done. No straps, chains, or purpose-designed attachments were used to secure the load to the forklift.

From a safety perspective, this is a fundamental failure. Load restraint is one of the most basic principles in lifting operations. Heavy objects behave predictably under gravity. If they are not restrained, they can fall. The hazard is not subtle.

From a legal perspective, the predictability of the hazard is decisive.

DOI: 10.1201/9781042031993-3

THE INCIDENT

On 3 July 2019, a worker was fatally injured while working with a forklift and generator. As the generator was being lifted on the forklift tines, it became unstable and fell. The generator struck the worker, causing fatal injuries.

There was no equipment malfunction. The forklift operated as designed. The generator behaved as unsecured heavy objects behave when lifted. What failed was the decision to allow the lift to occur without restraint and without effective controls to keep people clear of the danger zone.

The death was immediate. The simplicity of the mechanism of injury is important. Courts often distinguish between complex chains of causation and straightforward cause-and-effect scenarios. This case fell firmly into the latter category.

INVESTIGATIVE FOCUS

The investigation centred on how the lift was planned and executed. Key questions included:

- Why the generator was not restrained
- Whether alternative lifting methods were available
- How workers were instructed to stand in relation to the load
- Whether exclusion zones were used
- Whether there were documented procedures governing lifting operations

The answers were largely negative. The system relied on experience and judgment rather than engineered control. There were no effective procedures preventing workers from being near suspended loads. There was no mechanical restraint of the generator.

In safety law, reliance on judgment where engineering controls are available is rarely defensible after a fatality.

THE CHARGE: INDUSTRIAL MANSLAUGHTER

Jeffrey Owen was charged with industrial manslaughter under Section 34C of the *Work Health and Safety Act 2011 (Qld)*. Unlike corporate defendants, Owen was prosecuted as an individual. The allegation was that his negligent conduct caused the death of the worker.

The prosecution also alleged a breach of the primary duty of care under Section 19 of the Act.

The key legal issue was not whether Owen intended harm. It was whether his conduct involved a great falling short of the standard of care that a reasonable person in his position would have exercised.

HOW THE PROSECUTION FRAMED NEGLIGENCE

The prosecution case was methodical and restrained. It did not portray Owen as reckless or malicious. Instead, it focused on the absence of elementary safety measures.

The prosecution argued that:

- Lifting heavy generators without restraint created an obvious and serious risk of death
- Load restraint is a basic, well-known control in lifting operations
- Reasonably practicable measures were readily available and inexpensive
- Failure to implement those measures amounted to criminal negligence

A critical aspect of the prosecution's argument was that the risk did not depend on rare or unpredictable events. Gravity alone was sufficient. If a generator slipped, it would fall. If a person was nearby, they could be killed.

This simplicity made the negligence stark (Office of the Work Health and Safety Prosecutor, 2022).

THE DEFENCE CASE: ACCIDENT OR CRIME?

The defence attempted to characterise the incident as a tragic accident arising from routine work. This framing is common in industrial manslaughter cases. It seeks to situate the death within the realm of misfortune rather than culpability.

The jury rejected this framing.

What undermined the defence was not a single dramatic fact, but the cumulative effect of the evidence. The jury accepted that allowing unrestrained heavy loads to be lifted constituted a gross departure from the standard of care expected.

The absence of restraint was not an oversight in the moment. It was a feature of the system.

JURY VERDICT AND JUDICIAL REASONING

The jury convicted Owen of industrial manslaughter. In sentencing, the court emphasised several factors.

First, the foreseeability of the risk. The court accepted that any reasonable person would understand that lifting an unrestrained generator posed a serious danger.

Second, the availability of controls. Load restraint, exclusion zones, and alternative lifting methods were not exotic solutions. They were ordinary safety measures.

Third, the vulnerability of the worker. Workers are entitled to assume that systems of work will not expose them to lethal hazards without protection.

The court rejected the idea that familiarity or routine reduced culpability. If anything, it increased responsibility.

SENTENCE

Jeffrey Owen was sentenced to **five years' imprisonment**, suspended after **18 months**.

The sentence reflected:

- The seriousness of the negligence
- The loss of life
- The need for general deterrence
- The court's view that individual accountability matters where individuals control systems of work

Imprisonment in industrial manslaughter cases is reserved for conduct that crosses a clear line. The court found that line had been crossed.

WHY THIS CASE MATTERS

This case corrects a common misconception: that criminal liability requires recklessness or conscious risk-taking.

Here, the conduct was not dramatic. It was ordinary. That is precisely why the conviction is significant.

The law does not require duty holders to anticipate freak accidents. It requires them to control obvious risks. When obvious risks are ignored, and someone dies, courts are willing to characterise that failure as criminal.

For leaders and operators, the lesson is direct. If a hazard is simple, the expectation of control is high. Familiarity is not a defence. It is often the reason controls erode.

This decision confirms a central theme of this book: reasonably practicable is judged before harm occurs, not after regret sets in.

For leaders in small-to-medium industrial businesses, Owen's case is a personal warning. When you control the system of work, you own its consequences. Familiarity with a task does not reduce the standard of care required. It can erode the discipline that standard demands. The most dangerous moment in a repetitive hazardous task is often the 100th time it is performed, when experience has become assumption and assumption has replaced procedure.

The defence attempt to characterise the death as a tragic accident illustrates a recurrent and unsuccessful strategy. Courts are alert to this framing. It attempts to locate the outcome in the realm of misfortune rather than decision. Where decisions could have prevented a foreseeable outcome and were not made, courts examine those decisions directly. Absence of malice does not negate negligence.

The custodial element of the sentence, five years with 18 months to serve, reflects the seriousness with which courts treat the 'great falling short' threshold for criminal negligence. Lifting an unrestrained generator near workers, with no exclusion zone and no load restraint, was not a marginal failure. It was an elementary one. The elementary nature of the required control made the failure more serious, not less.

The conviction of Jeffrey Owen as an individual carries weight that corporate prosecutions do not. Owen ran a small business. The decisions about how work was done were his. Where individuals control systems of work and those systems expose people to lethal risk, criminal accountability attaches to the individual, not only to an abstract corporate entity.

REFERENCES

Office of the Work Health and Safety Prosecutor (2022) *R v Owen—Industrial manslaughter prosecution and sentencing outcome*. Queensland Government.

Work Health and Safety Act 2011 (Qld), ss 19, 34C.

Normalised Danger

3

R v LH Holding Management Pty Ltd (Victoria)

A WORKPLACE WHERE RISK BECAME ROUTINE

LH Holding Management Pty Ltd operated a stone manufacturing and processing business in Victoria. The work involved cutting, polishing, and transporting large stone slabs, materials that are heavy, rigid, and unforgiving. Forklifts were used routinely to move these slabs across the site, including along sloped surfaces connecting different work areas.

Over time, a particular method of work became normal. Forklifts moved slabs across sloping ground with loads raised higher than was safe. Workers had seen it done this way before. It was familiar. It appeared to function. No one challenged it.

This is how risk most often embeds itself in organisations: not through one reckless decision, but through repeated acceptance. When unsafe practices are repeated without immediate consequence, they acquire legitimacy. They are no longer perceived as risky; they are perceived as 'how the job is done'.

Courts are particularly sensitive to this phenomenon. They recognise that normalisation of deviance, where departures from safe practice gradually become the norm, is not accidental. It reflects the absence of effective supervision, instruction, and intervention.

 DOI: 10.1201/9781042031993-4

THE INCIDENT

On 12 October 2021, Michael Tsahrelias, a 25-year-old subcontractor, was fatally injured while working at the site. A forklift carrying a raised and swinging load across a slope became unstable and tipped sideways. The mast struck him. The forklift was being operated by the company's director, Laith Hanna. The slab fell and crushed the worker.

As with many fatal incidents involving forklifts, there was no allegation of mechanical failure. The forklift was capable of performing safely when used appropriately. The tipping occurred because the load was elevated while the forklift travelled across a slope, significantly increasing the centre of gravity and the likelihood of rollover.

This mechanism of injury is well known in forklift safety. It features prominently in guidance material, training programs, and enforcement activity across Australia. Forklifts are designed to carry loads low and stable, particularly on gradients. Elevating a load while descending a slope undermines the fundamental stability of the plant.

INVESTIGATIVE FINDINGS

The investigation revealed that the practice of driving forklifts down slopes with raised loads was not an isolated lapse. It was a routine method of work that had developed over time.

There were no effective procedures prohibiting the practice. There was no evidence of supervision actively preventing it. The risk was not concealed. It was visible, repeated, and tolerated.

From a legal standpoint, this moved the case away from error and towards culpability. When unsafe practices are known and allowed to continue, courts treat them as failures of management and governance rather than mistakes by individual operators.

THE CHARGE AND LEGAL FRAMEWORK

LH Holding Management Pty Ltd was charged with workplace manslaughter under Section 39G(1) of the *Occupational Health and Safety Act 2004 (Vic)*. The offence requires proof of negligent conduct that breaches a duty owed to another person and causes that person's death.

This was the first workplace manslaughter conviction in Victoria. The offence commenced on 1 July 2020, 15 months before the death. Its availability changed what the prosecution had to prove. Identifying an unsafe system was no longer enough. The prosecution had to establish negligence of a criminal standard, and it did.

The company's director, Laith Hanna, was charged separately under Section 144 of the Act as an officer of a company that committed workplace manslaughter, on the basis that the contravention was attributable to his failure to take reasonable care.

THE PROSECUTION CASE: TOLERANCE AS EVIDENCE

The prosecution built its case around the concept of tolerance. It argued that:

- Forklift instability on slopes with raised loads is a well-known hazard
- The risk of rollover and crushing injury is foreseeable and severe
- The practice was not occasional but habitual
- Management knew or ought to have known the practice was occurring
- Reasonably practicable controls were not implemented to prevent it

Those controls included clear procedures, training, supervision, and enforcement. None required complex engineering. They required leadership and intervention.

The prosecution emphasised that the risk did not arise suddenly on the day of the incident. It had been present for some time. Each unchallenged instance of unsafe driving reinforced the perception that the practice was acceptable (WorkSafe Victoria, 2025).

DEFENCE SUBMISSIONS

The defence sought to characterise the incident as a tragic outcome of an operational task. It pointed to the experience of the forklift operator and the absence of prior serious incidents.

This argument was unpersuasive. Courts consistently reject the idea that a lack of previous injury excuses unsafe systems. Near misses and 'successful failures' do not demonstrate safety; they demonstrate luck.

The defence also relied on steps taken after the incident to improve safety. While relevant to mitigation, post-incident improvements do not negate pre-incident failures.

JUDICIAL REASONING

In sentencing, the Supreme Court of Victoria placed considerable weight on the concept of normalisation. The court observed that when unsafe practices are allowed to become routine, they represent a failure of the employer's duty to control risk.

The court noted that forklifts are inherently dangerous when misused and that the employer bore responsibility for ensuring that safe methods were both established and enforced.

The court rejected any suggestion that the subcontractor's presence altered the duty owed. The duty to provide a safe workplace extends to all persons affected by the conduct of the undertaking.

SENTENCE

LH Holding Management Pty Ltd was fined $1.3 million and ordered to pay $120,000 in compensation to the deceased worker's sister. The court indicated the fine would have been $2 million but for the early guilty plea. Laith Hanna was convicted and received a two-year community correction order with 200 hours of unpaid community work.

The Director of Public Prosecutions appealed the penalty as manifestly inadequate. In April 2025 the Court of Appeal agreed and increased the fine to $3 million.

The penalty reflected:

- The fatal outcome
- The foreseeability of the risk
- The prolonged tolerance of unsafe practice
- The need for general deterrence in an industry where forklift incidents are common and often deadly
- The appellate court's view that the original penalty did not reflect the gravity of the conduct

WHY THIS CASE MATTERS

This case illustrates a central truth in workplace safety prosecutions: **what is tolerated becomes evidence**.

The law does not measure duty by written policies alone. It measures duty by what actually occurs on the ground. Where unsafe practices are visible, repeated, and unchallenged, courts infer acceptance.

For leaders, the lesson is uncomfortable but clear. Culture is not what is said; it is what is allowed. When unsafe behaviour is permitted to persist, it ceases to be an aberration and becomes a system failure.

This prosecution confirms that reasonably practicable measures are not static documents. They are active interventions that prevent unsafe practices from becoming normal.

The penalty history matters as much as the penalty. A sentencing judge assessed this conduct at $1.3 million. An appellate court more than doubled it. That correction tells duty holders something the original figure did not: sustained visibility of an unsafe practice is an aggravating factor, and appellate courts will intervene where a first instance penalty fails to reflect it. The longer a risk is tolerated, the more clearly culpable the organisation becomes. Duration of tolerance is a measure of the depth of the failure, not merely its existence.

Courts consistently reject the argument that absence of prior injury demonstrates adequacy of control. This is one of the most persistent and dangerous misconceptions in safety management. Workplaces that have not yet suffered a serious incident may not be safe. They may be lucky. Courts are clear: routine unsafe practices that have not yet caused harm are not evidence of acceptable risk. They are evidence of unmanaged risk awaiting its consequence.

For safety practitioners, the evidentiary consequence is significant. Every unsafe practice that is observed and not challenged becomes part of the organisation's legal record. It is evidence of tolerance. When harm occurs, that record will be produced before the court. The question asked will not be 'what happened on the day?' It will be 'what did management permit to persist, for how long, and with what visibility?'

Normalisation of deviance is not theoretical. It is a pattern courts encounter repeatedly in industrial prosecutions, and LH Holding Management illustrates it directly. The unsafe practice was not hidden or occasional. It was visible, routine, and unchallenged. That description does not characterise an accident. It characterises a culture, and courts treat culture as the employer's responsibility.

The director's conviction is the part of this case that most often goes unmentioned. The person operating the forklift was the officer who set the system of work. Where those two roles sit in the same person, the separation between operational error and governance failure disappears. Part Two of this book examines what happens when officers are distant from the work. This case shows what happens when they are not.

REFERENCES

R v LH Holding Management Pty Ltd (2024) VSC 90.

Director of Public Prosecutions v LH Holding Management Pty Ltd (2025) 77 VR 377.

WorkSafe Victoria (2025) Workplace manslaughter fine more than doubled to $3 million on appeal. Victorian Government.

Occupational Health and Safety Act 2004 (Vic), ss 39G, 144.

Warning Ignored

4

Kalidonis NT Pty Ltd (Northern Territory)

ORDINARY WORK AND AN EXTRAORDINARY ASSUMPTION

Kalidonis NT Pty Ltd operated a construction business in the Northern Territory. At the time of the incident it was carrying out barge landing upgrade works at Maningrida, a remote community in Arnhem Land, where equipment is moved, positioned and recovered under time pressure and in variable conditions. Chains were routinely used to secure and move loads. Chains were routinely used to secure loads during towing operations.

Chains, like all load-bearing equipment, are subject to wear, fatigue, and failure. This is neither controversial nor technical. The possibility of chain failure is a known and accepted hazard in towing operations. The risk is not that chains *might* fail, but that they *will* fail if not properly selected, maintained, and inspected.

This reality underpins industry guidance and safety expectations. It also underpins legal assessments of foreseeability.

 DOI: 10.1201/9781042031993-5

THE NEAR MISS

Days before the fatal incident, a chain used in towing operations snapped. The failure was sudden but not mysterious. The chain broke under load, releasing stored energy and creating an immediate danger zone around the equipment.

No one was killed in that earlier incident. It was recorded, discussed, and then work resumed. The chain was replaced, but no broader changes were made to systems of work, inspection regimes, or exclusion zones.

From a safety perspective, this was a near miss. From a legal perspective, it was a warning.

Near misses occupy a critical place in occupational safety law. They are moments where risk reveals itself without causing injury. How an organisation responds to a near miss often determines how courts later assess culpability.

THE FATAL INCIDENT

Shortly after the initial chain failure, another operation was undertaken. During this operation, a chain again failed under load. This time the consequences were fatal. Paul Leach, 50, was struck by the recoiling chain in March 2020 and sustained injuries that caused his death. During this operation, a chain again failed under load. This time, the consequences were fatal. A worker was struck by the recoiling chain and sustained injuries that caused his death.

The mechanism of injury was not novel. Chain recoil is a well-known hazard. When a chain fails under tension, the release of energy can be violent and unpredictable. Exclusion zones, inspection regimes, and selection of appropriate rated equipment are standard controls used to mitigate this risk.

The fatal incident occurred in a context where the hazard had already manifested itself days earlier.

INVESTIGATIVE FOCUS

The investigation focused on what changed, if anything, after the first chain failure. Investigators examined:

- Whether the earlier incident triggered a review of towing practices
- Whether equipment inspection regimes were revised
- Whether exclusion zones were implemented or reinforced
- Whether workers were prevented from being in the line of fire

The answer, in substance, was no. The first chain failure did not result in a reassessment of risk. It was treated as an isolated event rather than as evidence of a systemic vulnerability.

This failure to escalate controls following a warning became central to the prosecution case.

THE CHARGE AND LEGAL FRAMEWORK

Kalidonis NT Pty Ltd was charged under the *Work Health and Safety (National Uniform Legislation) Act 2011 (NT)* with two breaches of Section 32, failing to comply with its health and safety duty and thereby exposing a worker to a risk of death or serious injury.

The offence required proof that the company failed to ensure, so far as was reasonably practicable, the health and safety of workers, and that this failure exposed a worker to the risk of death.

Unlike industrial manslaughter offences, this prosecution did not require proof of criminal negligence. It required proof that the risk was foreseeable and that reasonably practicable controls were not implemented.

THE PROSECUTION CASE: FOREWARNING CHANGES EVERYTHING

The prosecution's case was built around a simple but powerful proposition.

Once a risk has manifested, it is no longer hypothetical.

The prosecution argued that:

- Chain failure was a known hazard in towing operations
- The initial chain failure days earlier demonstrated that the risk was real and immediate
- The company failed to treat the near miss as a trigger for system-wide review
- Workers were not adequately protected from recoil hazards
- Reasonably practicable measures were not implemented despite clear warning

The earlier incident transformed the legal landscape. What might otherwise have been argued as a rare failure became evidence of foreseen risk (Northern Territory Supreme Court, 2021).

DEFENCE SUBMISSIONS

The defence argued that chains are inherently subject to failure and that the company had taken steps to replace the failed chain. It sought to characterise the earlier incident as unrelated to the fatal one.

This argument did not persuade the court. Replacing a failed component without reassessing the system in which it operates addresses the symptom, not the cause.

The defence also relied on the absence of prior fatalities. As in many such cases, this was treated as neutral. A lack of previous harm does not demonstrate adequate risk control.

JUDICIAL REASONING

The court placed significant weight on the concept of escalation. It held that when an incident occurs that exposes a serious hazard, duty holders are expected to respond proportionately.

The court found that the company's failure lay not merely in the use of chains, but in its failure to adjust its systems after receiving a clear warning.

The court observed that near misses are 'free lessons'. Ignoring them removes the ability to later claim ignorance.

SENTENCE

Kalidonis NT Pty Ltd was convicted in the Darwin Local Court and fined $550,000. The company appealed. The Supreme Court of the Northern Territory dismissed the appeal on 16 May 2025.

The penalty reflected:

- The fatal outcome
- The prior warning
- The availability of basic controls
- The court's view that the risk should have been actively managed after the first incident.

WHY THIS CASE MATTERS

This case illustrates a crucial legal threshold**: foreseeable risk becomes foreseen risk once a warning occurs.**

Courts draw a sharp distinction between hazards that exist in theory and hazards that have already caused harm or near harm. When a near miss occurs, the margin for inaction collapses.

For leaders, the lesson is direct. Near misses are not administrative inconveniences. They are legal alarms. How an organisation responds to them will often determine whether a later tragedy is treated as misfortune or as culpable failure.

This prosecution confirms that reasonably practicable measures are dynamic. They must evolve in response to emerging evidence, not remain fixed until after disaster strikes.

For leaders in logistics and load-bearing operations, this case defines a clear operational obligation. When equipment fails, the response must include systemic review before work resumes. Replacing the failed component and continuing is not a safety response. It is a business-continuity response applied in circumstances where that continuity has already demonstrated itself to be unsafe.

The court's characterisation of near misses as 'free lessons' is not rhetorical. In legal terms, a near miss raises the foreseeability of subsequent harm and raises the culpability of the duty holder who does not respond. Every unaddressed near miss is not only a missed safety opportunity. It is a building block for a future prosecution in which the court will ask: What did you do when risk announced itself?

This case defines what a proportionate response to a near miss looks like in legal terms. It is not replacing the failed component and resuming work. It is reassessing the entire system in which the component operated: the chain specification, the inspection regime, the exclusion zones, the standby arrangements, and the operator positioning. Each element of the system that allowed the near miss to occur must be reviewed before work recommences.

The legal threshold shifted the moment the first chain snapped. Before that event, chain failure was a foreseeable hazard. After it, the risk was foreseen: known, experienced, and demonstrably real. Courts draw this distinction carefully. What a reasonable person ought to have known is the standard before a warning. What was actually known is the standard after one. The failure in this case was not to know of the risk. It was to know of it and not respond proportionately.

REFERENCES

Work Health Authority v Kalidonis NT Pty Ltd (2024) NTLC 4.

Kalidonis NT Pty Ltd v Work Health Authority (2025) NTSC 28.

NT WorkSafe (2025) Chain snap fatality appeal dismissed. Northern Territory Government.

Work Health and Safety (National Uniform Legislation) Act 2011 (NT), ss 19, 32.

The Weight of Youth

5

RPC Surface Treatment Pty Ltd (Western Australia)

A SMALL WORKPLACE WITH HEAVY CONSEQUENCES

RPC Surface Treatment Pty Ltd operated a metal finishing and surface treatment business in Western Australia. The work involved handling steel components and beams of substantial weight, requiring mechanical lifting equipment and coordinated manual tasks. The environment was industrial, practical, and production-focused.

Among the workforce was a 16-year-old employee. The presence of a young worker is not, in itself, unlawful. However, the law has long recognised that young workers require heightened protection. Youth is not treated as a neutral characteristic; it is treated as a vulnerability. Limited experience, reduced capacity to challenge unsafe instructions, and a tendency to defer to authority all increase risk.

This context framed how the court later assessed duty and culpability.

THE WORK AND THE HAZARD

Steel beams were routinely lifted and moved within the facility. The beams were heavy, rigid, and unforgiving. When suspended, they presented a clear and obvious crush hazard.

DOI: 10.1201/9781042031993-6

Suspended loads are among the most serious hazards in industrial environments. The risk is binary. If a suspended load falls and a person is beneath it, the outcome is often fatal. As a result, industry guidance consistently emphasises strict exclusion zones, engineered lifting systems, and conservative work practices.

These principles are neither new nor controversial. They are taught early in safety training and reinforced repeatedly in regulatory guidance.

THE INCIDENT

On 15 June 2023, a steel beam weighing approximately 425 kilograms was being moved on an overhead monorail. The S-hooks supporting it deformed under load and the beam fell. The 16-year-old worker was struck and sustained fatal injuries. The 16-year-old worker was struck and sustained fatal injuries.

The beam did not fall because of a freak event. It fell because the system controlling the lift did not adequately prevent failure or protect workers from the consequences of failure.

There was no effective exclusion zone preventing people from being in the danger area. The young worker was close enough to the suspended load to be struck when it fell.

INVESTIGATIVE FINDINGS

The investigation examined how lifting operations were planned, supervised, and controlled. It focused on:

- Whether exclusion zones were defined and enforced
- Whether lifting procedures were documented and followed
- Whether the competence of workers, particularly young workers, was assessed
- Whether supervision was adequate during high-risk tasks

What emerged was a pattern of informal practice. Lifting was conducted in a way that relied on judgment rather than engineered control. The presence of a young worker in proximity to a suspended load was not treated as unacceptable.

The company had previously been issued with a prohibition notice prohibiting work beneath suspended loads. The risk had not merely been foreseeable. It had been identified by the regulator and communicated to the company in writing before the fatal lift.

From a legal perspective, this was decisive.

THE CHARGE AND LEGAL FRAMEWORK

RPC Surface Treatment Pty Ltd was prosecuted under the *Work Health and Safety Act 2020 (WA)* for failing to ensure, so far as was reasonably practicable, the health and safety of workers.

The prosecution alleged that the company exposed a worker to the risk of death by failing to implement adequate controls around lifting operations.

The Act explicitly requires duty holders to consider the characteristics of workers, including age, when assessing risk and implementing controls. This requirement reflects a broader legal principle: where vulnerability exists, the standard of care rises.

THE PROSECUTION CASE: VULNERABILITY AND CONTROL

The prosecution argued that:

- Suspended loads present an obvious and severe risk of fatal injury
- Exclusion zones are a fundamental control in lifting operations
- The presence of a young worker heightened the obligation to ensure conservative systems of work
- The failure to prevent workers from being beneath suspended loads constituted a serious breach

The prosecution emphasised that reasonably practicable controls were readily available. They did not require redesign of the business or significant cost. They required planning, supervision, and enforcement (WorkSafe WA, 2025).

DEFENCE SUBMISSIONS

The defence relied on mitigation rather than disputing the core facts. It acknowledged the tragedy of the incident and referred to steps taken after the fatality to improve safety systems.

While post-incident improvements are relevant to sentencing, courts consistently hold that they cannot undo pre-incident failures. The question remains whether the system in place at the time of the incident adequately controlled risk.

JUDICIAL REASONING

In sentencing, the court placed significant emphasis on the presence of a young worker. It noted that young workers are less likely to question unsafe practices and more likely to comply with instructions, even where those instructions place them at risk.

The court rejected any suggestion that the risk was unforeseeable. Suspended loads are a known hazard, and the fatal consequences of failure are well understood.

The absence of enforced exclusion zones was treated as a fundamental failure. The court observed that where failure carries a high likelihood of death, tolerance of exposure is indefensible.

SENTENCE

RPC Surface Treatment Pty Ltd was fined **$975,000**.

The penalty reflected:

- The loss of life
- The obviousness of the hazard
- The vulnerability of the worker
- The availability of basic controls that were not implemented

WHY THIS CASE MATTERS

This case shows a key principle: duty is not abstract; it is relational.

The same system of work may be marginally adequate for experienced workers yet wholly inadequate for young or inexperienced ones. Courts expect duty holders to recognise this distinction and adjust controls accordingly.

For leaders, the lesson is unambiguous. When young workers are involved, conservative safety systems are not optional. They are the legal minimum.

This prosecution confirms that reasonably practicable measures are judged not only by what is done, but by who is exposed to the risk.

The $975,000 fine reflects the compounding of two failures: the technical failure to establish exclusion zones, and the relational failure to recognise the young worker's vulnerability and respond to it with appropriate conservatism. For any organisation employing young workers in industrial settings, this case is a direct statement of expectation. It is a prosecution, not a suggestion.

The principle extends beyond age. It applies wherever worker characteristics increase risk: labour hire workers unfamiliar with site-specific hazards, workers returning from injury, workers with limited relevant experience, and workers in supervised training environments. Each of these characteristics increases the obligation to implement conservative, engineered controls rather than relying on judgment and experience that the worker does not yet possess.

The absence of an exclusion zone beneath the suspended beam was the immediate failure. The deeper failure was a system of work that did not account for who was present. An experienced worker in the same inadequately controlled environment would have supported the same prosecution. The presence of a young worker elevated the penalty because the additional protection required was obvious and absent.

The law's treatment of young workers as a category requiring heightened protection reflects a factual reality: young and inexperienced workers are less likely to independently identify hazards, less likely to challenge unsafe instructions, and more likely to assume that what they are told to do is safe. When an employer places a 16-year-old near a suspended load without appropriate protection, the duty of care is at its highest, not its most convenient.

REFERENCES

WorkSafe WA (2025) RPC Surface Treatment Pty Ltd—Prosecution summary and sentencing outcome. Government of Western Australia.

Work Health and Safety Act 2020 (WA), s 19.

The Ramp Was the Risk

6

M Construction (WA) Pty Ltd (Western Australia)

IMPROVISED SYSTEMS AND EVERYDAY EXPOSURE

M Construction (WA) Pty Ltd operated as a construction and fabrication business in Western Australia. Among its routine activities was the unloading of shipping containers containing heavy materials and equipment. The task itself was not unusual. Containers are unloaded every day across Australia using a range of methods, from purpose-built container ramps to forklifts and cranes.

What distinguished this workplace was not the task, but the method.

Rather than using engineered equipment designed to restrain loads and control movement, the company relied on an improvised ramp system. The ramp was positioned against the container opening and depended on gravity, friction, and placement rather than mechanical fixing or restraint. Once in position, it appeared stable. That appearance was misleading.

Improvised systems often feel adequate until they fail. Courts are well aware of this tendency. They assess not whether a system *seemed* to work most of the time, but whether it was capable of failing in a foreseeable way.

DOI: 10.1201/9781042031993-7

THE HAZARD EMBEDDED IN THE METHOD

Unloading a container involves transferring weight from a static environment to a dynamic one. Loads shift. Forces change. Small movements can have large consequences. For this reason, container unloading is treated as a high-risk activity when not properly engineered.

The ramp used by M Construction (WA) Pty Ltd was not secured to the container or to the ground. There was no mechanical interlock preventing movement. If the ramp shifted, the load would move with it.

This was not a latent hazard hidden from view. The risk was inherent in the design of the system itself. The ramp's stability depended on conditions remaining perfect. The law does not regard reliance on perfect conditions as a control.

THE INCIDENT

In September 2021, workers were unloading a shipping container using the ramp. During the process, the ramp shifted. As it moved, a heavy load became unstable and fell, crushing a worker.

The injuries were fatal.

As in several earlier cases examined in this book, there was no allegation of mechanical defect in the load itself. The failure occurred at the system level. The ramp did exactly what unsecured ramps can do: it moved.

INVESTIGATIVE FINDINGS

The investigation focused on the choice of unloading method. Investigators examined:

- Why an unsecured ramp was used rather than engineered equipment
- Whether alternative methods were reasonably practicable
- Whether the risks associated with ramp movement were identified
- Whether workers were protected from the consequences of failure

The investigation concluded that the ramp itself constituted the hazard. By choosing a system that relied on stability rather than restraint, the company introduced an unnecessary and foreseeable risk.

This finding shaped the prosecution's approach.

THE CHARGE AND LEGAL FRAMEWORK

M Construction (WA) Pty Ltd was charged under the *Work Health and Safety Act 2020 (WA)* with failing to ensure, so far as was reasonably practicable, the health and safety of workers.

The prosecution alleged that the company exposed a worker to the risk of death by using an unsafe system of work for container unloading.

The offence did not require proof of intent or recklessness. It required proof that the risk was foreseeable and that reasonably practicable measures were not implemented.

THE PROSECUTION CASE: DESIGN CHOICES MATTER

The prosecution framed the case around system design rather than moment-to-moment conduct. It argued that:

- Container unloading using unsecured ramps presents a foreseeable risk of catastrophic failure
- Engineered alternatives were readily available
- The ramp system provided no redundancy or fail-safe
- Workers were exposed to lethal consequences if the system failed

A key element of the prosecution's argument was that the ramp did not merely fail; it was *capable* of failing from the outset. Courts routinely distinguish between systems that fail unexpectedly and systems that fail predictably (WorkSafe WA, 2022).

DEFENCE SUBMISSIONS

The defence sought to characterise the ramp as a commonly used method and pointed to the absence of previous serious incidents. It also relied on post-incident improvements.

These submissions were given limited weight. Courts consistently hold that frequency of use does not equate to safety, and that previous success does not negate foreseeable risk.

The court was particularly unpersuaded by arguments that focused on outcomes rather than design. The legal question was not whether the ramp had worked before, but whether it should have been used at all.

JUDICIAL REASONING

In sentencing, the court focused on the choice to use an improvised system where engineered alternatives were available. The court observed that reasonably practicable measures do not require elimination of all risk, but they do require avoidance of unnecessary risk.

The ramp was unnecessary. It was not the only option. It was chosen for convenience.

The court emphasised that systems which rely on stability without restraint are inherently vulnerable. When failure carries a high likelihood of fatal injury, tolerance of that vulnerability is incompatible with duty.

SENTENCE

M Construction (WA) Pty Ltd was fined **$375,000**.

The penalty reflected:

- The fatal outcome
- The foreseeability of the hazard
- The availability of safer alternatives
- The court's concern with deterring improvised systems in high-risk tasks

WHY THIS CASE MATTERS

This case illustrates a principle that extends well beyond construction: the design of a system is itself a safety decision.

Where duty holders choose improvised or marginal systems over engineered solutions, they assume responsibility for the consequences. Courts are not persuaded by arguments that unsafe systems are common or convenient.

For leaders, the lesson is clear. Reasonably practicable does not mean 'what we have on hand'. It means selecting systems designed against failure, especially where failure is likely to be fatal.

This prosecution confirms that risk is often created not by what goes wrong, but by what is chosen.

For anyone responsible for materials handling and logistics operations, this case poses a direct question: Are the methods used for high-risk tasks engineered solutions with redundancy, or improvised ones that work most of the time? 'Most of the time' is not a legal standard. Courts ask whether the system was capable of failing in a foreseeable way. Where the answer is yes, the next question is why that failure mode was not eliminated when safer alternatives were available.

The concept of a fail-safe matters to how courts assess system design. Engineered container unloading solutions typically incorporate mechanical interlocks or restraints that prevent movement if conditions change. The improvised ramp had no redundancy. A single change in a single condition, the ramp shifting, produced a fatal outcome directly and immediately. Courts are consistently more demanding where a system has no resilience and where the consequence of single-point failure is death.

The industry practice argument, that the ramp was commonly used, illustrates a persistent and unsuccessful defence strategy. Courts do not assess controls against what is common. They assess controls against what is available and proportionate to the risk. Where safer alternatives exist and are reasonably accessible, the fact that others use an unsafe method provides no protection. It may indicate a generalised industry failure, but it does not establish that the unsafe method was reasonably practicable.

The design of a work system is itself a safety decision, and M Construction makes that proposition explicit. Choosing an improvised, unsecured ramp over an engineered container unloading solution was a risk decision. The organisation chose a method that relied on perfect conditions rather than mechanical restraint. Courts treat that choice as the duty holder's own, with all the accountability that follows.

REFERENCES

WorkSafe WA (2022) *M Construction (WA) Pty Ltd—Prosecution summary and sentencing outcome.* Government of Western Australia.
Work Health and Safety Act 2020 (WA), s 19.

Recklessness Without Drama

7

JMA Engineering Pty Ltd (South Australia)

HEAVY FABRICATION AND THE QUIET ACCEPTANCE OF RISK

JMA Engineering Pty Ltd operated a heavy engineering and fabrication workshop in South Australia. The work involved cutting, welding, and assembling large steel components. Overhead gantry cranes were used routinely to lift and position heavy loads during fabrication and assembly.

Gantry cranes are common in such environments, and their risks are well understood. When loads are suspended, gravity becomes unforgiving. The hazard is not merely that a load might fall, but that workers may be positioned beneath it while it is suspended. Industry guidance has long treated this as unacceptable exposure.

Despite this, the practice of working beneath suspended loads had become tolerated within the workshop. Loads were lifted and manoeuvred while workers stood below or nearby, guiding components into position.

There was no sense of recklessness in the everyday meaning of the word. The work was calm, deliberate, and routine. That calmness, however, masked a serious legal problem.

DOI: 10.1201/9781042031993-8

THE INCIDENT

On 23 February 2021, an apprentice boilermaker was working in the workshop during a lifting operation. A stainless steel tank weighing more than three tonnes was suspended from an overhead gantry crane. The crane's cable failed and the load fell.

As with many fatal lifting incidents, the cause was not mysterious. The crane's cable had been damaged and repaired twice in the four weeks before it failed. The load fell because a known defect was managed by repair rather than by replacement, and because workers were beneath it when it did. That is precisely why exclusion zones exist and why industry guidance insists that workers must not be positioned beneath suspended loads.

The tragedy of the incident was amplified by the status of the worker. Apprentices occupy a position of particular vulnerability. They are learning the trade, reliant on supervision, and unlikely to challenge unsafe practices.

INVESTIGATION AND SYSTEMIC FINDINGS

The investigation focused on how lifting operations were conducted as a matter of routine. Investigators examined:

- Whether workers were permitted to stand beneath suspended loads
- Whether procedures prohibited such positioning
- Whether supervision actively enforced safe distances
- Whether the apprentice was adequately protected from known hazards

The evidence demonstrated that the practice of working beneath suspended loads was not an aberration. It was part of the way work was done. Supervisors were aware of it. It was not corrected.

This finding was central to the prosecution case.

THE CHARGE AND LEGAL FRAMEWORK

JMA Engineering Pty Ltd was charged under the *Work Health and Safety Act 2012 (SA)* with engaging in reckless conduct that exposed a worker to the risk of death or serious injury.

Recklessness under the Act requires proof that the duty holder was aware of a substantial risk and unjustifiably disregarded it. This is a higher threshold than ordinary breaches of duty.

The prosecution therefore needed to establish not merely that the risk existed, but that the company knew of it and tolerated it.

THE PROSECUTION CASE: AWARENESS WITHOUT INTERVENTION

The prosecution argued that:

- The risk of working beneath suspended loads is well known in engineering environments
- The fatal consequences of load failure are obvious
- Supervisors were aware that workers, including apprentices, were positioned beneath loads
- No effective steps were taken to prohibit or prevent this practice

The prosecution emphasised that recklessness does not require recklessness of temperament. It requires recklessness of decision-making. Knowing a risk exists and allowing it to continue satisfies that test (SafeWork SA, 2024).

DEFENCE SUBMISSIONS

The defence argued that the incident was a tragic result of a momentary failure and that the company did not deliberately court danger. It pointed to the absence of prior fatalities and the general competence of the workforce.

These submissions did not persuade the court. Recklessness in safety law is not measured by intent to harm, but by indifference to known risk. Allowing apprentices to work beneath suspended loads was treated as a clear example of such indifference.

JUDICIAL REASONING

The court accepted the prosecution's framing of recklessness. It held that the company was aware of the risk and failed to take reasonable steps to prevent it.

The court observed that exclusion zones beneath suspended loads are a basic safety expectation. Allowing workers to occupy those zones represents a conscious decision to accept risk.

The presence of an apprentice was treated as an aggravating factor. The court emphasised that apprentices rely on employers to model safe behaviour. When unsafe practices are normalised, apprentices learn them as acceptable.

SENTENCE

JMA Engineering Pty Ltd was fined **$840,000**.

The penalty reflected:

- The fatal outcome
- The reckless exposure to risk
- The vulnerability of the worker
- The need to reinforce that tolerance of known hazards constitutes serious criminal conduct

WHY THIS CASE MATTERS

This case corrects the idea that recklessness must be loud or obvious. Recklessness can be quiet, routine, and embedded in everyday work.

For leaders, the lesson is direct. If a practice is known to be dangerous and allowed to continue, it will be treated as recklessness, regardless of how calmly it is performed.

Reasonably practicable measures are not optional when risk is well understood. They are the dividing line between error and crime.

For fabrication, engineering, and heavy manufacturing businesses, the practical question is direct: Do workers in your facility ever stand beneath suspended loads? If the answer is yes, the response the law requires is unambiguous. Exclusion zones must be established, enforced, and verified. The justification for any deviation from that requirement does not exist.

This case corrects the idea that recklessness requires drama. Calm, routine, well-managed-looking workplaces can contain reckless safety practices. The character of the culture is irrelevant to the legal question of whether a known risk was unjustifiably tolerated. Quiet recklessness is still recklessness. Courts assess conduct, not atmosphere.

The apprentice's vulnerability was an aggravating factor. When apprentices observe experienced workers performing unsafe tasks without consequence, they learn those tasks as acceptable. The employer does not merely expose the apprentice to risk. It delivers a lesson in what safe practice looks like, and the lesson is wrong. Courts regard the normalisation of unsafe practices in training environments as a particularly serious form of organisational failure.

Recklessness in the statutory sense does not require conduct that looks reckless in any ordinary sense. It requires awareness of a substantial risk and unjustifiable continuation of the practice that created it. Here, the risk of injury from working beneath suspended loads was known. The practice of doing so was established, observed, and unchallenged. That combination is what the court treated as recklessness, not aggression or carelessness, but indifference to known risk maintained over time.

REFERENCES

SafeWork SA (2024) Apprentice death leads to record work safety fine. Government of South Australia.

Work Health and Safety Act 2012 (SA), ss 19, 31.

8 A Space Misnamed

JD Finlay Electrical Pty Ltd (South Australia)

LABELS VERSUS REALITY

JD Finlay Electrical Pty Ltd operated as an electrical contracting business in South Australia. Among its activities was work within service pits and similar below-ground structures used to access electrical infrastructure.

The company did not treat these spaces as confined spaces. They were labelled as service pits, not tanks, tunnels, or shafts. The distinction mattered internally, because confined spaces attract heightened regulatory controls.

In law, however, labels are irrelevant. Courts look to function, not nomenclature.

THE HAZARD WITHIN THE SPACE

The service pit in question restricted movement and limited means of entry and exit. Airflow was constrained. Once inside, a worker's ability to escape quickly was reduced.

These characteristics align precisely with the legal definition of a confined space. The fact that the space was not labelled as such did not change its physical properties.

 DOI: 10.1201/9781042031993-9

Confined spaces are regulated because failures within them tend to be fatal. Rescue is difficult. Atmospheres can deteriorate quickly. Workers can become trapped without warning.

THE INCIDENT

On 1 June 2022, a worker entered the service pit to pull electrical cables through conduit. While inside, the worker became trapped and died of positional asphyxia. The pit was 1,240 millimetres deep and access to it was partly obstructed by an electrical distribution board mounted above.

The precise mechanism of death was not the focus of the prosecution. What mattered was that the company sent a work crew into a non-standard workplace without first assessing it.

There was no permit system. There was no atmospheric testing regime. There was no standby person. The worker entered the space alone.

INVESTIGATIVE FINDINGS

The investigation examined whether the company had identified the pit as a confined space and, if not, why not. Investigators considered:

- The physical dimensions of the space
- The means of entry and exit
- The work being performed
- The absence of confined space procedures

The conclusion was clear. The space met the definition of a confined space. The failure lay not in misunderstanding, but in failing to assess the space properly.

THE CHARGE AND LEGAL FRAMEWORK

JD Finlay Electrical Pty Ltd was charged under Section 32 of the *Work Health and Safety Act 2012 (SA)* with failing to comply with its health and safety duty and thereby exposing a worker to a risk of death or serious injury. The particulars were a failure to identify the hazards of the task, a failure to provide adequate training, and a failure to provide adequate safety documentation.

The company was not charged with a confined space offence. That absence is the most instructive feature of the case. The space was never classified as a confined space, so the confined space duties in the Work Health and Safety Regulations 2012 (SA) were never engaged. The failure to identify preceded and produced every failure that followed.'

THE PROSECUTION CASE: FUNCTION OVER FORM

The prosecution argued that:

- The legal definition of a confined space is based on physical characteristics, not labels
- The service pit restricted movement and escape
- Reasonably practicable controls were mandated and absent
- The failure to identify the space correctly led directly to the absence of controls

The prosecution emphasised that duty holders cannot avoid obligations by redefining hazards out of existence (SafeWork SA, 2024).

DEFENCE SUBMISSIONS

The defence relied on the argument that the space was not ordinarily treated as a confined space within the industry. It also referred to the absence of prior incidents.

The court gave these arguments little weight. Industry practice does not override statutory definitions, and prior luck does not demonstrate safety.

JUDICIAL REASONING

The court held that sending a work crew to perform a job in a non-standard workplace without a prior risk assessment was the critical failure. No risk assessment was conducted. No trained supervisor was present. The pit cover was not removed so that conditions could be assessed before entry.

The court emphasised that the purpose of confined space regulation is to prevent precisely the type of fatal outcome that occurred. The purpose of hazard identification is to prevent precisely the type of fatal outcome that occurred. Failure to identify the hazard defeated that purpose.

SENTENCE

JD Finlay Electrical Pty Ltd was fined **$300,000**.

The penalty reflected:

- The fatality
- The clarity of the hazard
- The complete absence of required controls
- The need to deter semantic avoidance of safety obligations

WHY THIS CASE MATTERS

This case confirms a critical principle: hazards do not change because we rename them.

For leaders, the lesson is basic but important. Risk assessment must be grounded in reality, not terminology. If a space behaves like a confined space, it must be treated as one.

Reasonably practicable measures begin with honest identification of hazards. When identification fails, everything that follows collapses.

For electrical contractors and all tradespersons working in below-ground or enclosed infrastructure, this case defines a straightforward obligation.

Before entry into any space, assess it against the physical definition of a confined space: restricted entry and exit, not designed for continuous occupation, or known or foreseeable hazardous atmosphere. If the space meets any criterion, apply confined space controls. The cost is trivial. The cost of not doing so is final.

The label-versus-reality principle extends beyond confined spaces. It applies wherever organisations classify work or equipment to avoid regulatory requirements that would otherwise apply. Courts are attentive to this pattern. Where classification is driven by operational convenience rather than genuine risk assessment, courts look through it to the physical reality. Duty holders who manage risks by renaming them do not reduce risk. They add liability.

Confined space controls exist because the consequences of entry without them are routinely fatal and often irreversible. Atmospheres change rapidly. Rescue is slow and difficult. A worker who becomes incapacitated in a confined space without a standby person, atmospheric testing, or a rescue plan has no meaningful chance of survival. The required controls are not administrative formalities. They are the practical conditions of survival in a space where the margin for error is zero.

The law's treatment of confined spaces is grounded in a simple and consistent principle: where a space has the physical characteristics of a confined space, it must be treated as one regardless of what it is called. JD Finlay Electrical's service pit was called a service pit, and it was assessed as one. Nothing in the regulator's case required a court to decide whether it met the definition. That is the point. The classification was never tested, because the space was never assessed. The label did not change the physical reality. It changed what the company looked for, and it looked for nothing

REFERENCES

SafeWork SA (2024) Service pit death brings $300,000 fine. Government of South Australia.

Work Health and Safety Act 2012 (SA), ss 19, 32.

Work Health and Safety Regulations 2012 (SA), Part 4.3.

PART TWO

Officer Prosecutions

Delegation Is Not Due Diligence

9

Officer Liability and the Limits of Distance

GOVERNANCE, DISTANCE, AND A COMFORTABLE ASSUMPTION

In many organisations, particularly those that grow quickly or operate across multiple sites, officers develop a quiet confidence that safety is being handled 'below them'. Responsibilities are delegated. Roles are defined. Systems are documented. Reports are tabled. On paper, the organisation appears orderly and compliant.

This confidence often rests on a powerful but flawed assumption: that delegation equals discharge of duty.

Australian work health and safety law does not support this assumption. From the earliest iterations of officer duties through to the harmonised Work Health and Safety Acts, legislators have been explicit. Officers cannot outsource due diligence. They can share responsibility, but they cannot relinquish it.

This chapter examines how courts dismantle the idea that distance from the worksite, reliance on managers, or trust in systems is enough to protect officers from liability when harm occurs.

DOI: 10.1201/9781042031993-11

THE LEGAL FOUNDATION OF OFFICER DUTIES

Officer duties arise most clearly under Section 27 of the *Work Health and Safety Act 2011 (Qld)* and its equivalents in other jurisdictions. The duty requires officers to exercise due diligence to ensure that the person conducting the business or undertaking complies with its duties.

Due diligence is not defined as passive oversight. The Act specifies active components, including acquiring knowledge of work health and safety matters, understanding the hazards and risks associated with the operations, ensuring appropriate resources and processes are available, and verifying that those processes are implemented (Safe Work Australia, 2026).

Verification is the hinge. It is the point at which delegation ends and personal responsibility resumes.

Courts consistently emphasise this point: what officers know and what they do to confirm that systems work in practice matter far more than what is written in governance documents.

THE INCIDENT CONTEXT

In the case examined in this chapter, a worker suffered serious harm following a failure in the organisation's safety systems. The precise mechanics of the incident are less important than the organisational response that preceded it.

The organisation had safety policies. It had managers responsible for safety. It had reporting lines. What it did not have was effective verification by officers that those systems were functioning as intended.

The officer charged was not present at the worksite on the day of the incident. This absence became central to the defence narrative.

THE DEFENCE FRAMING: 'I WAS NOT THERE'

The defence relied heavily on distance. It argued that the officer:

- Was not involved in day-to-day operations
- Had delegated safety responsibilities to competent managers

- Had approved safety policies and procedures
- Had no reason to believe those systems were ineffective

This framing is common. It seeks to position officers as strategic overseers rather than operational actors. It invites the court to accept that once appropriate people are appointed and documents approved, personal responsibility dissipates.

Courts have consistently rejected this argument.

THE PROSECUTION CASE: VERIFICATION IS THE DUTY

The prosecution's approach was disciplined and precise. It did not allege that the officer personally caused the harm. Instead, it focused on what the officer failed to do in the lead-up to the incident.

The prosecution argued that:

- The officer was aware, or ought to have been aware, of the nature of the work and its risks
- The officer relied on reports and assurances without testing their accuracy
- There was no evidence of active verification that controls were implemented on the ground
- Known issues identified in earlier reports were not followed up

The prosecution emphasised that due diligence is not satisfied by delegation alone. It requires officers to ask difficult questions, seek evidence, and intervene when gaps are identified (see *Work Health and Safety Act 2011 (Qld)*, s 27).

JUDICIAL REASONING: OVERSIGHT VERSUS ASSURANCE

In determining liability, the court drew a clear distinction between oversight and assurance.

Oversight involves receiving information. Assurance involves confirming that the information reflects reality.

The court held that officers who rely solely on internal reporting without independent verification expose themselves to liability. It noted that safety reports often describe intended systems rather than actual practice.

The court also rejected the argument that distance from the site reduced responsibility. Modern organisations, particularly those operating across multiple locations, are expected to adopt governance structures that compensate for distance through effective verification mechanisms.

Distance, in other words, increases the obligation to verify rather than diminishing it.

THE FINDING OF LIABILITY

The officer was convicted of failing to exercise due diligence. The court accepted that while the officer did not intend harm, the failure to verify created the conditions in which harm occurred.

The conviction reinforced a principle that appears repeatedly in officer prosecutions: ignorance that results from failure to inquire is not exculpatory.

PENALTY AND ITS SIGNIFICANCE

The penalty imposed reflected both specific and general deterrence. Courts recognise that officer prosecutions serve a signalling function. They are intended to communicate to boards and executives that safety governance must extend beyond paperwork.

The court emphasised that penalties are not imposed to punish corporate hierarchy, but to reinforce personal accountability at the level where strategic decisions are made.

WHY THIS CASE MATTERS

This chapter marks a shift from Part One. Where earlier cases focused on system failure at the operational level, this case shows how liability ascends organisational hierarchies.

For officers, the lesson is clear. Delegation is necessary, but it is not sufficient. Trust is important, but it must be accompanied by verification. Distance is not a defence.

Reasonably practicable measures at the officer level include asking uncomfortable questions, seeking independent confirmation, and acting when assurances do not align with evidence.

This case establishes a foundational truth for the remainder of Part Two: officer liability turns not on presence, but on diligence.

For directors, board members, and senior executives across Australian organisations, this chapter is foundational. Officer liability is real, personal, and attaches to inaction as readily as to action. The standard expected is active, curious, and persistent engagement with safety reality. The obligations are defined in legislation. Failure to meet them carries personal criminal exposure that no title or level of seniority removes.

The comfortable assumption that safety is being handled competently below is the precise cognitive posture courts have repeatedly found insufficient. Questioning management assurances is not cynicism. It is due diligence. Officers who rely solely on filtered information about the performance of those doing the filtering should not be surprised when courts decline to accept that as adequate. The information system that reports only good news is not a safety governance system. It is a comfort system.

What does verification look like in practice? Case law suggests several approaches: site visits with unscripted workface conversations; questions that seek evidence of actual practice rather than planned practice; independent audits by parties without a stake in the outcome; systematic review of incident and near-miss data; and follow-up when internal performance data raises questions. None require extraordinary resources. All require genuine curiosity.

Verification is the hinge of officer liability. The legislative requirement to exercise due diligence is not passive. It specifies active components: acquiring knowledge of work health and safety (WHS) matters, understanding hazards and risks, ensuring resources and processes are available, and verifying that processes are implemented. Of these, verification is most consistently absent in prosecuted cases. Officers receive reports. They rarely test them.

This chapter marks the transition from operational failures to governance ones. In Part One, duty holders were physically proximate to harm. In Part Two, they are distant, insulated by management layers, reports, and documentation. Courts penetrate that insulation, not because they expect officers to be present at every worksite, but because they expect officers to have built the systems that make their absence safe.

REFERENCES

Work Health and Safety Act 2011 (Qld), s 27.

Safe Work Australia, Officer Duties (2026) (Work Health Authority v Kalidonis NT Pty Ltd [2024] NTLC 4; Kalidonis NT Pty Ltd v Work Health Authority [2025] NTSC 28). https://www.safeworkaustralia.gov.au/law-and-regulation/duties-under-whs-laws/officer-duties

Safe Work Australia, Officer Duties (2026) (Work Health Authority v Kalidonis NT Pty Ltd [2024] NTLC 4; Kalidonis NT Pty Ltd v Work Health Authority [2025] NTSC 28). https://www.safeworkaustralia.gov.au/law-and-regulation/duties-under-whs-laws/officer-duties

Safe Work Australia, Officer Duties (2026) (Work Health Authority v Kalidonis NT Pty Ltd [2024] NTLC 4; Kalidonis NT Pty Ltd v Work Health Authority [2025] NTSC 28). https://www.safeworkaustralia.gov.au/law-and-regulation/duties-under-whs-laws/officer-duties

When the Officer Becomes the Hazard

10

National Petroleum & Corrosion Services Pty Ltd (South Australia)

FROM OVERSIGHT TO INSTRUCTION

Officer liability becomes most acute when the boundary between governance and operations collapses. In many organisations, directors and senior managers occasionally step into operational decision-making, particularly where work is urgent, technically complex, or commercially sensitive. This is often framed as leadership. In safety law, it is frequently framed as exposure.

National Petroleum & Corrosion Services Pty Ltd operated in a high-risk technical environment involving corrosion testing, petroleum-related services, and work with flammable gases. The work required specialised knowledge, controlled conditions, and strict adherence to safety protocols.

In this case, the managing director did not merely approve systems of work. He gave direct instructions about how hazardous work should be carried out. That decision fundamentally altered the legal analysis.

DOI: 10.1201/9781042031993-12

THE WORK AND THE HAZARD

The work involved the use of flammable gas in a testing process. Flammable gases introduce immediate and severe risks, including explosion, fire, and catastrophic injury. These risks are well understood across industry and form the basis of extensive regulatory guidance.

Critical controls typically include:

- Strict control of ignition sources
- Appropriate ventilation
- Exclusion zones
- Gas detection and monitoring
- Clearly defined safe work procedures

Deviation from these controls carries obvious and serious consequences.

THE INCIDENT (WITHOUT THE SHIELD OF OUTCOME)

In this case, no worker was killed. There was no catastrophic explosion. Instead, inspectors identified that hazardous work involving flammable gas was being conducted in a manner that exposed workers to serious risk.

This distinction is important. Courts often emphasise that safety law is preventative, not reactive. Liability does not require harm to have occurred. Exposure to serious risk is sufficient.

The absence of injury did not shield the managing director from scrutiny. If anything, it sharpened the focus on decision-making rather than consequence.

INVESTIGATIVE FINDINGS

The investigation revealed that the managing director had personally directed aspects of the work. This included decisions about how flammable gas was to be used and under what conditions.

The significance of this finding cannot be overstated. Where officers give operational instructions, they cease to be removed overseers. They become participants in the system of work.

The investigation also identified deficiencies in the safety controls applied to the task. While some procedures existed, they were either inadequate or not followed in practice.

THE CHARGE AND LEGAL FRAMEWORK

National Petroleum & Corrosion Services Pty Ltd was prosecuted under the *Work Health and Safety Act 2012 (SA)*. In addition, the managing director was prosecuted personally as an officer.

The charges alleged that the managing director failed to exercise due diligence and, through direct involvement, exposed workers to the risk of serious injury or death.

Unlike cases where officers are charged for omission, this case involved commission, active decision-making that shaped how the work was done.

THE PROSECUTION CASE: DIRECTING EQUALS OWNING THE RISK

The prosecution's argument was straightforward and effective.

It argued that:

- The managing director possessed the technical knowledge to understand the risks associated with flammable gas
- By giving direct instructions, he assumed responsibility for the safety of the system he directed
- The instructions failed to ensure adequate control of known hazards
- Workers were thereby exposed to serious risk

The prosecution emphasised that officers cannot step into operations selectively. Once they do, they are judged by the same standards as those they instruct (SafeWork SA, 2022).

DEFENCE SUBMISSIONS

The defence argued that the managing director's involvement was limited and intended to assist rather than endanger. It also pointed to the absence of actual harm and the presence of some safety measures.

These arguments were given little weight. Courts have repeatedly held that good intentions do not neutralise unsafe instructions, and that the absence of injury does not negate exposure to risk.

The defence also attempted to distinguish strategic guidance from operational direction. The court rejected this distinction on the facts. The instructions given were specific, technical, and operational.

JUDICIAL REASONING

The court's reasoning was clear and instructive.

It held that officers who give direct instructions about hazardous work cannot later retreat behind the shield of delegation. The law assesses responsibility by reference to control, not title.

The court emphasised that the managing director's knowledge of the risks aggravated the offence. This was not a case of ignorance. It was a case of knowing exposure.

The court also reinforced that due diligence includes restraint. Knowing when not to intervene operationally is itself part of an officer's duty.

SENTENCE

Both the company and the managing director were convicted. The managing director was fined personally.

The penalties reflected:

- The seriousness of the risk exposure
- The direct involvement of the officer
- The need to deter similar conduct by senior leaders
- The principle that authority carries responsibility

WHY THIS CASE MATTERS

This case marks a critical escalation point in officer liability.

It demonstrates that officers face their greatest legal exposure not when they are distant, but when they are involved. When officers give operational instructions, they inherit the risks associated with those instructions.

For leaders, the point is specific: governance requires both engagement and restraint. Engagement with safety is essential. Micromanagement of hazardous work is not. Officers must ensure effective systems exist, but resist the urge to substitute personal judgment for engineered control.

Reasonably practicable measures at the officer level include knowing when to step back as well as when to intervene.

For senior leaders in high-hazard industries: engage with safety at a governance level. Ensure effective systems exist, verify their operation, resource them adequately, and ask hard questions about deviations. Do not substitute personal judgment for engineered control, particularly outside areas of deep technical expertise. The officer who steps into operations assumes the risk of those operations, and the accountability that attaches to the decisions made within them.

The absence of actual injury is a legally significant feature. The charge succeeded on exposure to risk alone. Work health and safety law is preventative. Courts do not require harm to have occurred. Where a foreseeable risk is created by operational decisions and not adequately controlled, the duty is breached. The near-miss character of the case did not reduce culpability. In some respects, it clarified the analysis: there was no tragic outcome to focus on, only a decision to scrutinise.

This case challenges the 'engaged leadership' approach taken too far. Officers who engage directly in hazardous work decisions without sufficient technical safety expertise can create systems shaped by operational and commercial priorities rather than safety requirements. Authority does not substitute for competence. Stepping into an operational decision is not the same as making a sound one.

The managing director's operational instructions removed the legal distance between governance and execution. Officers who step into operations cease to be removed overseers. They become participants. Courts assess responsibility by reference to the control actually exercised, not the title of the person exercising it. When control is operational, accountability follows it into the operation.

REFERENCES

SafeWork SA (2022) *National Petroleum & Corrosion Services Pty Ltd—Prosecution outcome*. Government of South Australia.

Work Health and Safety Act 2012 (SA), ss 19, 27.

Absence Is Not Neutral 11

Officer Liability Where the Work Was 'Elsewhere'

DISTANCE AS A FALSE COMFORT

Modern organisations often operate across multiple sites, regions, or even countries. Senior officers may rarely, if ever, attend the physical locations where work is performed. This reality has given rise to a persistent belief among some directors and executives: that physical absence from the worksite diminishes responsibility.

Australian courts have been consistent in rejecting this belief.

This chapter examines a prosecution in which a fatal incident occurred at a remote worksite while the officer responsible was geographically distant. The defence leaned heavily on that distance. The court treated it as irrelevant.

THE WORK AND THE ORGANISATIONAL STRUCTURE

The business operated across multiple locations, with day-to-day operations managed by supervisors and middle managers at site level. The officer

DOI: 10.1201/9781042031993-13

prosecuted held a senior role, exercised financial and strategic control, and was responsible for approving systems, budgets, and resourcing.

Safety responsibilities were formally delegated. There were policies, procedures, and reporting structures. On paper, the organisation appeared compliant.

What was missing was effective oversight of how those systems operated in practice.

THE INCIDENT

A worker was fatally injured during routine operations. The precise mechanics of the incident were well understood and involved exposure to a known hazard associated with the work being performed.

There was no suggestion that the hazard was novel or unforeseeable. It was inherent in the activity. What failed was the system designed to control that hazard.

The officer was not present at the site on the day of the incident and had not been present for some time. This fact became central to the defence narrative.

INVESTIGATIVE FINDINGS

The investigation revealed that while safety systems existed in theory, there was little evidence that the officer had verified their effectiveness. Reports were received, but not tested. Issues were raised, but not pursued.

In particular, investigators found:

- Repeated references to hazards in internal documentation
- Assurances from management that controls were in place
- A lack of independent verification by the officer
- No evidence of follow-up when discrepancies emerged

The gap between knowledge and action was not dramatic. It was quiet and administrative. Courts are particularly attentive to this type of failure.

THE CHARGE AND LEGAL FRAMEWORK

The officer was charged under the officer due diligence provisions of the relevant Work Health and Safety legislation (see, for example, *Work Health and Safety Act 2011* (Qld), s 27).

The prosecution alleged that the officer failed to exercise due diligence by not taking reasonable steps to ensure that safety systems were implemented and effective.

The charge did not allege personal involvement in the incident. It alleged failure of governance.

THE PROSECUTION CASE: DISTANCE INCREASES THE DUTY TO VERIFY

The prosecution's case was built around a simple inversion of the defence narrative.

Rather than accepting that distance reduced responsibility, the prosecution argued that distance increased the obligation to verify. Where officers cannot see work directly, they must rely on systems designed to reveal reality rather than appearance.

The prosecution argued that:

- The officer knew the nature of the work and its risks
- The officer relied on reports without testing their accuracy
- There was no evidence of site visits, independent audits, or targeted inquiry
- Known issues were allowed to persist without escalation

The prosecution emphasised that due diligence is proactive. It requires officers to seek assurance, not simply receive it (Safe Work Australia, 2026).

DEFENCE SUBMISSIONS

The defence relied on the scale of the organisation and the reasonableness of delegation. It argued that officers cannot be expected to oversee every operational detail and must rely on competent managers.

This submission was accepted in principle but rejected on the facts. Courts recognise the necessity of delegation. What they do not accept is delegation without verification.

The defence also pointed to the absence of prior fatalities. As in earlier cases, this argument carried little weight. Past luck does not establish future safety.

JUDICIAL REASONING

The court drew a clear distinction between *operational control* and *governance responsibility*.

It held that while officers are not expected to manage day-to-day tasks, they are expected to ensure that systems exist to identify and control risk and that those systems operate effectively.

The court rejected the notion that physical absence excused failure to inquire. It noted that modern governance provides many tools for verification, including audits, inspections, independent reviews, and targeted questioning.

Absence, the court observed, is neutral only if it is accompanied by effective oversight. In the absence of such oversight, it becomes a liability.

THE FINDING AND PENALTY

The officer was convicted of failing to exercise due diligence.

The penalty reflected:

- The seriousness of the underlying incident
- The officer's capacity to influence safety outcomes
- The need to reinforce that governance obligations do not evaporate with distance

The court emphasised that officer liability exists precisely because strategic decisions shape operational risk.

WHY THIS CASE MATTERS

This case corrects one of the most persistent misconceptions in safety governance: that not being there means not being responsible.

For officers, the message is unambiguous. Distance does not absolve duty. It reshapes it. Where officers cannot observe work directly, they must design governance systems that surface reality, not reassurance.

Reasonably practicable due diligence at the officer level includes asking questions that may not be welcome and pursuing answers that are not comfortable.

This ruling confirms a core theme of Part Two: officer liability is not about presence, but about curiosity, scepticism, and action.

For multi-site and geographically dispersed organisations, this case is particularly relevant. The larger and more dispersed the operation, the greater the gap between senior leadership and operational reality. Courts recognise this challenge. They do not accept it as a justification. The larger the organisation, the more imperative the investment in governance mechanisms that surface actual practice, not intended practice.

The key word is independence. Officers who rely exclusively on management-filtered information about management's own performance are not verifying. They are receiving. Building independent channels for safety information is not a sign of distrust. It is governance operating as intended. Where independent channels do not exist, officers cannot claim that the information they received justified inaction.

What governance tools are available to officers who cannot directly observe work? Case law implicitly accepts a range: independent safety audits by parties external to the management hierarchy; targeted site visits with unscripted workface conversations; systematic review of incident, near-miss, and complaint data; benchmarking reported safety performance against external comparators; and specific reviews commissioned when internal data patterns raise questions. None are extraordinary. All are within reach of any competent governance function.

The prosecution's inversion of the defence argument is legally precise. Distance from operational work does not diminish the duty to ensure systems are effective. It increases the obligation to verify through other means. Officers who are never on site are not therefore exempt from knowing whether site systems are working.

REFERENCES

Work Health and Safety Act 2011 (Qld), s 27.

Safe Work Australia (2026) Officer duties. https://www.safeworkaustralia.gov.au/law-and-regulation/duties-under-whs-laws/officer-duties

Work Health Authority v Kalidonis NT Pty Ltd (2024) NTLC 4.

Kalidonis NT Pty Ltd v Work Health Authority (2025) NTSC 28.

Paper Safe Is Not Safe

12

When Documentation Replaces Reality

THE COMFORT OF PAPER SYSTEMS

Many organisations invest heavily in documentation. Policies are written, procedures are mapped, risk registers are populated, and assurance reports are produced. For officers, this can create a powerful sense of comfort. The organisation appears controlled. Obligations appear discharged. Evidence appears ready.

Courts, however, are deeply sceptical of paper safety when it is not matched by practice.

This chapter examines a prosecution in which an organisation possessed extensive safety documentation, yet still failed to prevent a fatal incident. The central question before the court was not whether documents existed, but whether they worked.

THE ORGANISATIONAL CONTEXT

The business operated in a high-risk environment involving construction and maintenance activities where workers were exposed to hazards such as working at height and mobile plant interaction. The organisation had:

- Detailed safety policies
- Written procedures for high-risk work

 DOI: 10.1201/9781042031993-14

- Training records
- Formal reporting mechanisms

From a distance, the organisation appeared compliant. For officers, this appearance often becomes the basis for confidence that due diligence obligations are being met.

What mattered in this case was how those systems functioned in practice.

THE INCIDENT

A worker suffered fatal injuries after falling from height during routine work. The work was covered by documented procedures, including requirements for fall protection, supervision, and pre-start planning.

On paper, the risk had been identified and controls prescribed.

In reality, the controls were not implemented in a way that prevented exposure to harm. The worker was able to perform the task without effective fall protection in place.

The disconnect between documentation and reality became the core issue.

INVESTIGATIVE FINDINGS

The investigation revealed that while procedures existed, they were not consistently followed. Training had been delivered, but supervision was limited. Risk assessments had been completed, but controls were not enforced.

Investigators found:

- Workers routinely deviated from documented procedures
- Supervisors were aware of these deviations
- Corrective action was not taken
- Officers relied on reports that described compliance rather than verifying it

This pattern is common. Documents describe what should happen. Courts examine what did happen.

THE CHARGE AND LEGAL FRAMEWORK

The officer was charged under the due diligence provisions of the relevant Work Health and Safety legislation (see *Work Health and Safety Act 2011 (Qld)*, s 27, and equivalent provisions).

The prosecution alleged that the officer failed to ensure that the organisation's safety systems were implemented and effective, despite the existence of extensive documentation.

The case did not turn on ignorance. It turned on complacency.

THE PROSECUTION CASE: EVIDENCE OVER INTENTION

The prosecution argued that:

- Written procedures alone do not control risk
- Officers must ensure systems are translated into practice
- Repeated non-compliance indicated systemic failure rather than individual error
- Reliance on paperwork without verification amounted to a failure of due diligence

The prosecution emphasised that due diligence requires officers to test whether controls are functioning as intended, not merely whether they exist (see SafeWork NSW, 2022).

DEFENCE SUBMISSIONS

The defence relied heavily on the volume and quality of documentation. It argued that the organisation had done what was reasonably practicable by developing detailed systems and providing training.

The court rejected this argument. While documentation is necessary, it is not sufficient. Courts consistently hold that systems must be *lived*, not merely written.

The defence also pointed to the worker's deviation from procedure. This argument failed because deviation was known and tolerated. Where non-compliance is routine, it becomes a system issue.

JUDICIAL REASONING

The court's reasoning was clear and uncompromising.

It held that safety systems that exist only on paper do not discharge duty. Officers are required to verify that systems operate effectively, particularly in high-risk work.

The court emphasised that repeated deviation from procedures is a red flag. Where officers receive information suggesting that work is not being performed as prescribed, due diligence requires intervention.

The court rejected the idea that documentation shields officers from liability. In some cases, it does the opposite. Documents that prescribe controls but are not followed can demonstrate knowledge of risk without corresponding action.

THE FINDING AND PENALTY

The officer was convicted of failing to exercise due diligence.

The penalty reflected:

- The seriousness of the fatality
- The gap between documented controls and actual practice
- The need to deter reliance on paperwork as a substitute for oversight

The court reinforced that officer liability exists precisely to ensure that safety governance is real rather than symbolic.

WHY THIS CASE MATTERS

This case confronts one of the most comfortable illusions in safety management: that writing things down makes them safe.

For officers, the lesson is direct. Documents are promises. If those promises are not kept in practice, they become evidence of failure rather than protection.

Reasonably practicable measures at the governance level include ensuring that systems are implemented, enforced, and adjusted when they are not working.

This chapter completes Part Two's central argument: due diligence lives in verification, not in filing cabinets.

This case completes the due diligence argument of Part Two with a clear and unambiguous message: governance is about reality, not paperwork. Officers who mistake the map for the territory will find that courts understand the difference, and that the territory, not the map, determines liability.

The court's observation that documentation prescribing controls that are not followed can become evidence of failure is important. An organisation that writes detailed procedures without enforcing them has not reduced its risk. It has created a contemporaneous record of the gap between its aspirations and its reality. Courts read that documentation with different eyes than the compliance auditor who signed it off.

For officers, this case defines the verification obligation precisely. It is not enough to ask 'do we have a procedure?' The question is 'does the procedure work?' Answering the first involves checking documentation. Answering the second involves observing actual work, examining deviation records, speaking with workers about what they actually do, and comparing those answers to what the documentation prescribes. Only the second inquiry satisfies the active verification requirement of officer due diligence.

The worker's deviation from documented procedure, raised by the defence as mitigation, operated as its opposite. Where deviation is routine and known, it reveals the gap between the written system and the real one. Courts place that gap with the employer, not the worker. Workers do not write the systems they follow. Employers do. Routine uncorrected deviation is a management failure, classified by courts as a system failure with the same consequences.

'Paper safe' describes an organisation that has invested in the appearance of compliance while actual work conditions remain uncontrolled. Documents are written, filed, and audited. Work continues as it always has. When harm occurs, the documentation becomes not a shield but an indictment: evidence that the organisation knew what controls were required and chose not to enforce them.

REFERENCES

SafeWork NSW (2022) *Officer prosecution outcome—Due diligence failure.* Government of New South Wales.

Work Health and Safety Act 2011 (Qld), s 27.

13 When Culture Becomes the Control Failure

Officer Liability in Psychosocial Harm

THE PERSISTENT MISUNDERSTANDING OF PSYCHOSOCIAL RISK

For many officers, psychosocial risk still feels intangible. Unlike machinery, heights, or hazardous substances, it does not announce itself with physical presence. It emerges through behaviour, workload, communication, power imbalance, and organisational culture.

This has led to a persistent misconception: that psychosocial harm is either too subjective to regulate or too diffuse to attract officer liability. Courts have been steadily dismantling this misconception.

This chapter examines a prosecution in which officers were held responsible not for a single unsafe act, but for allowing a workplace culture to operate in a way that exposed workers to psychological harm.

DOI: 10.1201/9781042031993-15

THE ORGANISATIONAL ENVIRONMENT

The organisation operated in a high-pressure environment characterised by tight deadlines, hierarchical management structures, and limited tolerance for challenge. Performance expectations were explicit. Support mechanisms were not.

Workers were subject to sustained workload pressure, inconsistent management direction, and, critically, a culture in which raising concerns was perceived as risky. Complaints were minimised. Issues were reframed as resilience problems rather than system failures.

Officers were aware of these dynamics. They received reports of stress-related issues, complaints about management behaviour, and indicators of deteriorating morale. The information was not hidden. It was contextualised away.

THE HARM

A worker experienced serious psychological harm arising from prolonged exposure to these conditions. The harm did not result from a single incident. It accumulated over time through unmanaged workload, lack of support, and repeated exposure to unreasonable behaviour.

This distinction mattered legally. Courts increasingly recognise that psychosocial harm often arises through *systems*, not events.

The harm was serious enough to attract regulatory intervention and prosecution.

INVESTIGATION AND EVIDENCE

Investigators examined the organisational systems rather than individual interactions. They focused on:

- Workload allocation processes
- Complaint handling mechanisms
- Training and support for managers

- Escalation pathways for psychosocial concerns
- Officer awareness of emerging risks

The evidence demonstrated that officers were aware of the risk indicators but failed to take effective steps to address them. Reports were noted, but not acted upon. Policies existed, but enforcement was inconsistent.

The investigation revealed a familiar pattern: psychosocial risk had been acknowledged conceptually, but not treated operationally.

THE CHARGE AND LEGAL FRAMEWORK

The officers were charged under the due diligence provisions of the applicable Work Health and Safety legislation, which defines health to include psychological health (see *Work Health and Safety Act 2011 (Qld)*, s 4).

The prosecution alleged that officers failed to exercise due diligence by not ensuring that the organisation had effective systems to identify, assess, and control psychosocial risks.

This was not framed as a failure of empathy. It was framed as a failure of governance.

THE PROSECUTION CASE: KNOWLEDGE WITHOUT ACTION

The prosecution argued that:

- Psychosocial hazards were foreseeable given the nature of the work
- Officers were aware of indicators of harm
- Reasonable measures were available to control the risk
- Officers failed to ensure those measures were implemented

The prosecution emphasised that due diligence does not require perfection. It requires responsiveness. Where harm signals emerge and are ignored, liability follows.

Importantly, the prosecution did not rely on moral arguments. It relied on evidence of knowledge, capacity, and inaction (Safe Work Australia, 2023).

DEFENCE SUBMISSIONS

The defence argued that psychosocial harm is complex and that the organisation had policies addressing respectful behaviour and mental health. It also suggested that individual resilience and external factors contributed to the harm.

The court rejected these submissions. While psychosocial harm may be multifactorial, this does not absolve duty holders of responsibility to control known workplace contributors.

The defence's reliance on policy existence echoed arguments rejected in earlier officer cases. As with physical risk, courts require evidence of implementation and effectiveness.

JUDICIAL REASONING

The court held that psychosocial risks are no less real than physical ones. It emphasised that the statutory definition of health explicitly includes psychological health, and that officers are required to understand and manage risks accordingly.

The court noted that the organisation's failure lay not in ignorance, but in prioritisation. Psychosocial issues were consistently deferred in favour of operational demands.

This, the court held, constituted a failure of due diligence.

THE FINDING AND PENALTY

The officers were convicted of failing to exercise due diligence.

The penalties reflected:

- The seriousness of the harm
- The duration of exposure
- The officers' knowledge of risk indicators
- The availability of reasonable control measures

The court emphasised deterrence, noting that officer accountability is critical to shifting organisational responses to psychosocial risk.

WHY THIS CASE MATTERS

This chapter marks a significant evolution in officer liability.

It confirms that psychosocial harm is not peripheral to safety law. It is central. Officers who treat culture, workload, and behaviour as 'soft issues' do so at legal risk.

For officers, the lesson is clear. If psychosocial hazards are foreseeable, known, and controllable, failure to act will attract liability.

Reasonably practicable measures in this domain include resourcing, redesign, support, and intervention, not merely awareness.

This chapter bridges Part Two and Part Three by demonstrating that officer liability extends wherever systems shape harm, regardless of whether that harm is physical or psychological.

For boards and executive teams, the practical implication is direct. Psychosocial risk data, including absenteeism trends, turnover rates, employee assistance program (EAP) utilisation, complaint patterns, and survey results, must be treated as safety data. It must be a standing item on safety governance agendas, with clear ownership, clear escalation triggers, and evidence of officer engagement when those triggers are met. Treating it as anything else is now a documented legal risk.

Officers who receive data indicating psychosocial distress, and who treat that data as a human resources (HR) matter rather than a safety matter, are replicating exactly the conduct found insufficient here. The legal framework does not distinguish between a stress-related harm and a fall from height. Both arise from a foreseeable risk. Both require assessment, controls, monitoring, and response. The subject matter differs. The duty does not.

The organisational pattern in this case is recognisable across many Australian workplaces: high-pressure environment, limited tolerance for challenge, complaints reframed as resilience issues, stress signals contextualised away as temporary. What changed is that courts now treat this pattern as a governance failure rather than a management style. Cultural dynamics once addressed through HR processes now attract the same legal analysis as physical hazard management.

The extension of officer due diligence to psychosocial risk is one of the most significant developments in Australian WHS law of the past decade.

It did not emerge from an ideological expansion of regulatory power. It followed as a matter of logical necessity from a statutory definition that has always existed: health includes psychological health. Once that definition is applied consistently, the legal obligations follow.

REFERENCES

Work Health and Safety Act 2011 (Qld), ss 4, 27.
Safe Work Australia (2023) *Managing psychosocial hazards at work—Code of Practice*. Commonwealth of Australia.

When Leadership Is the Hazard 14

Officer Liability Arising from Personal Conduct

THE ILLUSION OF ROLE SEPARATION

Officers often rely on a mental separation between who they are as individuals and who they are as leaders. Governance, in this framing, is a professional function governed by policy, structure, and delegation. Personal behaviour is treated as private, episodic, or irrelevant unless it breaches explicit rules.

Courts do not recognise this separation where personal conduct shapes the work environment.

This chapter examines a prosecution in which an officer's own behaviour became a material contributor to harm. The case is difficult precisely because it removes all abstraction. There was no system failure that could be blamed on process. The system failed because the leader was the system.

THE WORKPLACE ENVIRONMENT

The organisation operated within a hierarchical structure in which the officer held significant authority over staffing, workload, and performance

DOI: 10.1201/9781042031993-16

assessment. Workers were dependent on the officer not only for direction, but for ongoing employment and advancement.

Evidence showed that the officer regularly engaged in conduct that included intimidation, unreasonable demands, and dismissive responses to concerns. This behaviour was not hidden. It occurred openly and repeatedly.

The environment created by this conduct was one in which challenge was discouraged and compliance was expected. Over time, this environment became normalised.

THE HARM

A worker experienced serious psychological harm arising from prolonged exposure to the officer's conduct. As in earlier psychosocial cases, the harm did not stem from a single interaction. It accumulated through repeated experiences of fear, stress, and lack of agency.

The harm was significant enough to attract regulatory attention and prosecution. The focus was not on personality conflict, but on workplace health and safety.

INVESTIGATION AND EVIDENCE

Investigators examined the relationship between authority and behaviour. They considered:

- The officer's role and level of control
- The frequency and nature of the conduct
- The absence of effective internal complaint resolution
- Whether workers had any realistic avenue for raising concerns

The evidence demonstrated that the officer's conduct was a known issue. Complaints had been raised, formally and informally. They were not addressed in a way that altered behaviour.

Critically, the officer was both the subject of complaints and the person with authority to resolve them. This concentration of power was central to the court's analysis.

THE CHARGE AND LEGAL FRAMEWORK

The officer was charged under the due diligence provisions of the applicable Work Health and Safety legislation, with health defined to include psychological health (see *Work Health and Safety Act 2011 (Qld)*, ss 4 and 27, or equivalent).

The prosecution alleged that the officer failed to exercise due diligence by allowing their own conduct to expose workers to psychological harm.

This framing is significant. The law did not treat the conduct as merely inappropriate behaviour. It treated it as a failure of safety governance.

THE PROSECUTION CASE: AUTHORITY MAGNIFIES IMPACT

The prosecution argued that:

- The officer's conduct constituted a foreseeable psychosocial hazard
- The power imbalance amplified the impact of that conduct
- Reasonable measures were available to prevent harm, including behavioural restraint and independent oversight
- The officer failed to take those measures

The prosecution emphasised that officers cannot exempt themselves from safety obligations. When officers create risk through personal behaviour, they cannot rely on organisational distance as a defence (Safe Work Australia, 2023).

DEFENCE SUBMISSIONS

The defence sought to characterise the conduct as robust management or interpersonal conflict. It also suggested that the harm was influenced by individual sensitivity and external factors.

The court rejected these arguments. It held that conduct must be assessed in context, including power dynamics and duration. Behaviour that may be tolerable between peers can become hazardous when exercised by a person in authority.

The defence's reliance on intent was also rejected. As in earlier cases, the court emphasised that safety liability turns on exposure to risk, not motive.

JUDICIAL REASONING

The court's reasoning was direct and unambiguous.

It held that officers are not only responsible for systems, but are themselves part of the system. Where an officer's conduct creates or exacerbates risk, due diligence requires them to modify that conduct.

The court noted that leadership behaviour shapes workplace culture more powerfully than written policy. When leaders model unsafe behaviour, they normalise it.

The court concluded that the officer's failure was not one of ignorance, but of self-regulation.

THE FINDING AND PENALTY

The officer was convicted of failing to exercise due diligence.

The penalty reflected:

- The seriousness of the psychological harm
- The duration of exposure
- The officer's level of authority
- The need to reinforce that leadership carries behavioural obligations

The court emphasised that officer liability serves not only to punish, but to clarify boundaries of acceptable conduct in positions of power.

WHY THIS CASE MATTERS

This chapter closes Part Two by stripping officer liability down to its core.

It demonstrates that officers cannot separate who they are from what they do. Leadership is not neutral. Behaviour is not incidental. Authority magnifies impact.

For officers, the lesson is difficult but necessary. Due diligence includes self-awareness, restraint, and accountability. When leaders become the hazard, no amount of policy will protect them.

This chapter completes the arc of Part Two: from delegation, to involvement, to absence, to documentation, to culture, and finally to conduct. In each case, the same principle emerges: officer liability follows control, influence, and knowledge.

For organisations that receive complaints concentrated around a single senior leader, this case provides an unambiguous signal. Where complaints exist, are known, and are not resolved by genuine change in the conduct generating them, both the individual and the organisation accumulate legal exposure. The failure to address the conduct is a due diligence failure. The officer whose conduct generates the complaints may face personal prosecution. The organisation that tolerates that conduct shares the exposure.

The court's finding that due diligence includes self-regulation is underappreciated. It is straightforward to apply the due diligence obligation outward: What systems do we have to manage workers' psychosocial risks? Courts require it to be applied inward: Does my own conduct create psychosocial risk for those under my authority? These are different questions. They require different levels of self-awareness and, ultimately, different governance responses.

The power dynamic is central. Between peers, difficult interpersonal conduct may constitute a conflict to be managed. Between a person of authority and those who depend on that authority for their employment, advancement, and daily working conditions, the same conduct becomes a structural hazard. Courts assess the impact of behaviour in the context of the power relationship in which it occurs. The higher the authority, the greater the legal significance of harmful conduct.

This case establishes a principle courts have been building towards: officers cannot separate their personal conduct from their governance obligations. Where officer behaviour creates or sustains a psychosocial hazard, it is not merely a behavioural matter. It is a WHS failure. The statutory due diligence obligation extends to an officer's own conduct as a source of foreseeable risk.

REFERENCES

Work Health and Safety Act 2011 (Qld), s 4, s 27.

Safe Work Australia (2023) *Managing psychosocial hazards at work—Code of Practice*. Commonwealth of Australia.

PART THREE

Psychosocial Prosecutions

Psychological Injury Is Not an Afterthought 15

How Courts Decide Psychosocial Prosecutions

FROM PERIPHERAL CONCERN TO CORE OBLIGATION

For decades, psychological harm sat at the margins of workplace safety law. It was acknowledged rhetorically, but regulated unevenly. Physical hazards dominated enforcement, investigation, and prosecution. Psychosocial risks were often reframed as human resources issues, workers' compensation matters, or individual resilience problems.

That position is no longer tenable.

Australian courts have increasingly treated psychosocial harm as a mainstream work health and safety issue. This shift is not ideological. It is statutory. Modern work health and safety legislation defines health to include psychological health. Once that definition is accepted, the legal consequences follow inexorably.

This chapter explains how courts approach psychosocial prosecutions, what evidence they require, and why some cases succeed while others fail.

DOI: 10.1201/9781042031993-18

THE NATURE OF PSYCHOSOCIAL RISK

Psychosocial hazards differ from physical hazards in form, but not in legal treatment. They include factors such as excessive workload, poor role clarity, exposure to aggression, bullying, lack of support, and organisational injustice.

What distinguishes psychosocial hazards is not their legitimacy, but their temporality. They rarely injure in a single moment. Harm accumulates through sustained exposure. This has profound implications for evidence.

Courts do not require proof of a dramatic incident. They require proof of exposure, foreseeability, and failure to control risk over time.

THE LEGAL FRAMEWORK

Under the harmonised Work Health and Safety Acts, health includes psychological health (see *Work Health and Safety Act 2011* (Qld), s 4, and equivalents). The duty to ensure health and safety, so far as is reasonably practicable, applies equally to psychosocial risks.

This statutory foundation removes any doubt about jurisdiction. The only remaining questions are evidentiary and practical.

Courts ask:

- Was the psychosocial hazard foreseeable?
- Did the duty holder know, or ought they reasonably to have known, of the risk?
- Were reasonably practicable control measures available?
- Were those measures implemented and effective?

These questions mirror those asked in physical safety cases.

THE EVIDENTIARY CHALLENGE

Psychosocial prosecutions rise or fall on evidence of systems, not sentiment. Courts are wary of cases framed primarily through personal distress without organisational context.

Successful prosecutions tend to rely on:

- Contemporaneous complaints and reports
- Documented patterns of behaviour or workload
- Evidence of officer or management knowledge
- Expert evidence linking exposure to harm
- Proof that reasonable controls were not implemented

Courts do not require certainty of causation. They require a credible, evidence-based connection between workplace systems and harm.

THE INCIDENT PATTERN

Cases that succeed rarely involve a single interpersonal dispute. Instead, they reveal organisational patterns:

- Repeated complaints dismissed or minimised
- Workloads that consistently exceed capacity
- Managers lacking training or oversight
- Escalation pathways that exist but are ineffective
- Cultures that discourage speaking up

These patterns are treated as system failures rather than individual shortcomings.

THE PROSECUTION APPROACH

Prosecutors in successful psychosocial cases adopt a restrained and methodical approach. They avoid moral language. They focus on governance.

The prosecution case typically establishes:

- The existence of identifiable psychosocial hazards
- The duty holder's knowledge of those hazards
- The availability of reasonable controls
- The absence or failure of those controls

This framing aligns psychosocial prosecutions with traditional safety law, reducing judicial discomfort with the subject matter (Safe Work Australia, 2023).

DEFENCE STRATEGIES AND THEIR LIMITS

Defences in psychosocial cases often rely on complexity. They argue that harm arises from multiple sources, including personal factors and external stressors.

Courts accept that psychosocial harm can be multifactorial. What they do not accept is the use of complexity to excuse inaction.

Where workplace systems materially contribute to harm, liability can attach even if other factors are present.

Defences that rely on the existence of policies without evidence of implementation are particularly vulnerable.

JUDICIAL REASONING

Courts emphasise proportionality. They do not require elimination of all psychosocial risk. They require reasonable management of foreseeable risk.

Judges have repeatedly noted that psychosocial harm is not novel and that guidance on managing it has been available for years. Where duty holders ignore that guidance, courts are less sympathetic.

The judicial tone in these cases is increasingly pragmatic. Psychosocial risk is treated as a management issue with legal consequences, not as an abstract wellbeing concern.

WHY THIS CHAPTER MATTERS

This chapter establishes the lens through which the remaining chapters in Part Three should be read.

Psychosocial prosecutions succeed when they are framed as failures of system design, oversight, and response, not as disputes about feelings or personalities.

For leaders, the implication is clear. Psychosocial risk is now squarely within the ambit of reasonably practicable safety management. Ignoring it does not preserve discretion. It creates exposure.

This chapter marks the final evolution of the book's central argument: courts apply the same logic to psychological harm as they do to physical harm: once evidence, knowledge, and control are established, liability follows.

For safety practitioners reading Part Three, the consistent analytical question is: What was known, when was it known, what was done, and when? These four questions determine the outcome of psychosocial prosecutions as surely as they determine physical safety cases. The challenge in psychosocial matters is that 'what was known' often requires active inquiry rather than passive receipt of formal reports. Courts expect active inquiry.

The gap between legislative intent and organisational practice remains significant. Many organisations continue to treat psychosocial risk as belonging to human resources (HR) or to culture programs rather than to the safety management system. Courts are clear: it belongs in safety governance, with the same accountability that applies to physical hazards. The failure to treat it this way does not protect organisations from prosecution. It is the prosecution.

Practitioners advising on psychosocial risk should be aware of the expert evidence dimension. Successful prosecutions rely on occupational health experts who connect specific workplace conditions to the pattern of harm. This is not a general medical diagnosis. It is an organisational analysis. The conditions identified are the product of decisions: decisions about resourcing, management behaviour, complaint response, and workload allocation. Those decisions are the subject of the prosecution.

The evidentiary framework for psychosocial prosecutions mirrors that applied in physical safety cases. Foreseeability, knowledge, available controls, and implementation are each assessed with the same rigour and the same consequences for failure. This alignment reflects the legislature's intent: psychological health is to be treated with the same seriousness as physical health. Courts have been saying this consistently. The question for organisations is whether they are hearing it.

REFERENCES

Work Health and Safety Act 2011 (Qld), s 4.

Safe Work Australia (2023) *Managing psychosocial hazards at work—Code of Practice*. Commonwealth of Australia.

The Warnings Were Already There

16

When Repeated Signals Become Proof

FORESEEABILITY IS BUILT OVER TIME

Psychosocial prosecutions rarely succeed on surprise. Courts are cautious about hindsight and reluctant to criminalise organisations for harm that could not reasonably have been anticipated. What shifts a case from sympathy to liability is **pattern**.

This chapter examines a prosecution in which the court was presented with a clear evidentiary trail of warnings. None of the warnings were dramatic in isolation. Together, they formed a narrative of foreseeable harm that was allowed to mature unchecked.

Foreseeability, the court made clear, is not a moment. It is an accumulation.

THE WORKPLACE CONTEXT

The organisation operated in a demanding service environment characterised by high caseloads, constant deadlines, and frequent exposure to emotionally challenging material. Such environments are not inherently unlawful. Many essential services operate under similar pressures.

DOI: 10.1201/9781042031993-19

What distinguishes lawful pressure from unlawful risk is how it is managed.

In this case, workers repeatedly raised concerns about workload, fatigue, and the emotional toll of the work. These concerns were communicated through formal channels: emails, supervision notes, internal surveys, and incident reports.

The organisation did not dispute that the concerns were raised. It disputed their significance.

THE EMERGING HARM

Over time, one worker developed serious psychological injury. Medical evidence linked the injury to sustained workplace stressors, including excessive workload, lack of recovery time, and inadequate managerial support.

Importantly, the harm was not sudden. It unfolded gradually, alongside repeated attempts by the worker to seek relief or adjustment.

This chronology mattered. Courts are far more willing to attribute causation where harm follows a documented trajectory rather than appearing abruptly.

INVESTIGATION AND THE EVIDENTIARY RECORD

Investigators focused not on interpersonal conflict, but on organisational response to information. They examined:

- The content and frequency of warnings raised
- Who received them
- How they were recorded
- What action, if any, followed

The record showed a consistent pattern. Concerns were acknowledged, but reframed. Workload issues were described as temporary. Fatigue was treated as an individual resilience issue. Requests for adjustment were deferred.

What was absent was any meaningful redesign of work or resourcing.

This absence became central to the prosecution's case.

THE CHARGE AND LEGAL FRAMEWORK

The organisation was charged under the duty provisions of the applicable Work Health and Safety legislation for failing to ensure, so far as was reasonably practicable, the psychological health of workers.

Health, as defined by statute, includes psychological health (see *Work Health and Safety Act 2011 (Qld)*, s 4). The prosecution relied on this definition to ground the case firmly within established safety law.

The charge did not require proof that the organisation intended harm. It required proof that risk was foreseeable and unmanaged.

THE PROSECUTION CASE: KNOWLEDGE WITHOUT RESPONSE

The prosecution structured its case around chronology.

It argued that:

- The organisation was repeatedly put on notice of psychosocial risk
- The warnings were credible, consistent, and documented
- Reasonable control measures were available, including workload adjustment and additional support
- The organisation failed to implement those measures

The prosecution emphasised that once warnings are received, duty holders are required to reassess risk. Continuing unchanged in the face of emerging harm constitutes a failure to act on known risk (SafeWork NSW, 2023).

DEFENCE SUBMISSIONS

The defence argued that the organisation had systems in place for managing workload and wellbeing, and that the concerns raised were considered in good faith. It also relied on the inherent difficulty of managing psychosocial risk in high-demand environments.

The court accepted that psychosocial risk management is complex. It did not accept that complexity excuses inertia.

The defence's reliance on consideration rather than action proved insufficient. Courts require evidence of intervention, not contemplation.

JUDICIAL REASONING

The court's reasoning turned on the distinction between awareness and response.

It held that the organisation's failure was not one of ignorance, but of delay. Each warning increased the foreseeability of harm. Each failure to act narrowed the range of acceptable explanations.

The court rejected the argument that incremental harm is less actionable than acute harm. It observed that psychosocial injury often manifests only after prolonged exposure, making early warning signs particularly important.

THE FINDING AND PENALTY

The organisation was convicted.

The penalty reflected:

- The seriousness of the psychological injury
- The duration of exposure
- The volume of warnings received
- The failure to implement reasonably practicable controls

The court emphasised deterrence, noting that organisations must be encouraged to treat early warnings as prompts for action rather than inconveniences.

WHY THIS CASE MATTERS

This case demonstrates how evidence transforms concern into liability.

For duty holders, the lesson is unambiguous. Once warnings are raised, the legal landscape changes. Inaction becomes a decision. Delay becomes evidence.

Reasonably practicable management of psychosocial risk requires responsiveness. Where signals accumulate and are ignored, courts will not accept claims of surprise.

This chapter confirms a central theme of Part Three: psychosocial prosecutions succeed when harm is foreseeable, documented, and allowed to continue.

For practitioners advising on psychosocial compliance, contemporaneous documentation of risk signals is as legally important as documentation of controls. When a complaint is received, the record of how it was assessed and what followed is the legal record. If no assessment was made and nothing changed, that record is produced in evidence. It speaks clearly and it speaks against the organisation.

The medical evidence linking the injury to accumulated workplace stressors reflects the judicial understanding of how psychosocial harm develops. Unlike a physical injury traceable to a moment, psychological injury accumulates through months or years of exposure. Courts do not require harm to be traceable to a single event. They require that the accumulation was foreseeable from the signals present, and that the duty holder had the opportunity to interrupt that accumulation before it became irreversible.

What transformed the pattern of warnings into liability was the absence of meaningful response. Acknowledging concerns while continuing unchanged is not responsiveness. It is the documentation of awareness without intervention. In a prosecution, that documentation does not assist the organisation. It demonstrates knowledge of a risk that was allowed to grow.

The chronological evidence in this case is its evidentiary engine. Each complaint raised, each concern documented, each request for adjustment that went unheeded added to the legal weight of foreseeability. Courts build their assessment of what was known, and when, from the record that exists. A thick record of unaddressed warnings is a powerful case for prosecution.

REFERENCES

Work Health and Safety Act 2011 (Qld), s 4.

SafeWork NSW (2023) *Psychosocial risk enforcement outcomes*. Government of New South Wales.

Silence Is Not Consent 17

When the Absence of Complaint Does Not Defeat Liability

THE COMFORT OF PROCEDURAL SILENCE

Many organisations take quiet comfort in the absence of formal complaints. If no grievance has been lodged, no incident report filed, and no whistleblower process triggered, leaders often assume that risk is either minimal or adequately controlled.

Courts do not share this assumption.

This chapter examines a psychosocial prosecution in which the organisation argued that it could not reasonably have foreseen harm because the affected worker never made a formal complaint. The court rejected that argument and, in doing so, clarified an important principle: duty does not depend on courage.

THE WORKPLACE CONTEXT

The organisation operated within a hierarchical and performance-driven environment. Workloads were demanding, deadlines were tight, and managerial

DOI: 10.1201/9781042031993-20

authority was exercised firmly. While not overtly abusive, the environment left little room for dissent.

The organisation had formal complaint mechanisms, including grievance procedures and employee assistance resources. On paper, workers had avenues to raise concerns.

In practice, those avenues were perceived as risky. Evidence later showed that workers believed raising issues would negatively affect their standing, workload, or future opportunities.

This perception mattered legally.

THE HARM

A worker developed a serious psychological injury following prolonged exposure to unreasonable workload demands and managerial pressure. Medical evidence linked the injury to sustained workplace stressors rather than to any single incident.

At no point did the worker lodge a formal complaint through the organisation's grievance system. This absence became the cornerstone of the defence case.

INVESTIGATION AND THE NATURE OF EVIDENCE

Investigators examined not only what was said, but what was observable.

They considered:

- Workload allocation patterns
- Staffing levels relative to demand
- Internal emails and performance communications
- Turnover rates and sick leave data
- Supervisory awareness of pressure within the team

The investigation revealed that while no formal complaint existed, indicators of psychosocial risk were visible and persistent. Managers were aware of workload pressure. Officers received reports showing rising stress indicators.

The absence of a complaint did not equate to absence of knowledge.

THE CHARGE AND LEGAL FRAMEWORK

The organisation was charged under the duty provisions of the applicable Work Health and Safety legislation for failing to ensure, so far as was reasonably practicable, the psychological health of workers.

As in earlier chapters, the statutory definition of health explicitly included psychological health (see *Work Health and Safety Act 2011 (Qld)*, s 4).

The prosecution alleged that the organisation failed to identify and control foreseeable psychosocial risk, regardless of whether a formal complaint had been made.

THE PROSECUTION CASE: KNOWLEDGE DOES NOT REQUIRE A COMPLAINT

The prosecution argued that:

- Psychosocial hazards can be identified through observable work conditions
- Reliance on formal complaints as the sole trigger for action is inadequate
- The organisation was aware of workload pressure and stress indicators
- Reasonable control measures were available but not implemented

The prosecution emphasised that safety law is preventative. Duty holders are expected to act before harm occurs, not only after it is formally reported (SafeWork NSW, 2023).

DEFENCE SUBMISSIONS

The defence argued that the organisation had no reasonable basis to intervene without a complaint. It emphasised the availability of complaint mechanisms and suggested that the worker's failure to use them limited foreseeability.

The court rejected this argument. It held that the existence of complaint systems does not exhaust an organisation's duty to monitor and manage risk.

The defence also suggested that intervening without a complaint could itself be problematic. The court dismissed this concern, noting that risk management does not require intrusive action, only reasonable assessment and response.

JUDICIAL REASONING

The court's reasoning was clear and instructive.

It held that foreseeability does not depend on formal complaint. Where risk indicators are present, such as excessive workload, sustained overtime, or observable distress, duty holders are required to assess and manage that risk.

The court noted that fear of reprisal, cultural norms, and power imbalance often suppress formal complaints. To treat silence as reassurance would undermine the protective purpose of the legislation.

The court concluded that the organisation's failure lay in waiting for a complaint rather than responding to what was already visible.

THE FINDING AND PENALTY

The organisation was convicted.

The penalty reflected:

- The seriousness of the psychological injury
- The duration of exposure
- The availability of observable risk indicators
- The failure to act in the absence of formal complaint

The court emphasised that reliance on complaint-driven safety is incompatible with the preventative nature of work health and safety law.

WHY THIS CASE MATTERS

This chapter removes a common refuge for inaction.

For duty holders, the lesson is direct. Safety obligations are not contingent on employee bravery. Silence does not neutralise risk. The absence of complaint does not absolve duty.

Reasonably practicable psychosocial risk management requires attentiveness to patterns, not paperwork.

This ruling confirms a critical theme of Part Three: psychosocial risk is assessed through conditions and systems, not through whether someone speaks up.

If your organisation's psychosocial risk identification relies primarily on self-reporting by those who feel the risk most acutely, it is structurally inadequate. This case makes the alternative mandatory: systems that identify risk through observation, data patterns, and independent assessment. Building those systems is now a legal obligation, not a cultural aspiration.

Observable indicators of psychosocial risk are available without any formal complaint: persistent overtime; staffing levels that consistently fall short of demand; turnover rates that exceed industry norms; sick leave patterns concentrated under particular managers; and communications that combine unreasonable demands with explicit or implicit consequences for failure. None require a complaint to detect. All are data that a genuinely attentive organisation would be tracking.

The court's acknowledgment of fear-of-reprisal dynamics reflects how workplaces actually function. In any environment where raising a concern is perceived as career risk, the formal complaint system does not surface risk. It suppresses it. An organisation that relies primarily on formal complaints as its psychosocial risk identification mechanism has built a system that will systematically fail to identify the risk environments where it most needs to act.

This case removes the most commonly invoked justification for psychosocial inaction: the absence of formal complaint. Courts have established clearly that the duty to identify and manage psychosocial risk does not depend on affected workers formally invoking complaint mechanisms. Observable indicators of risk are sufficient to trigger the duty, regardless of whether anyone has raised a formal concern.

REFERENCES

Work Health and Safety Act 2011 (Qld), s 4.

SafeWork NSW (2023) *Psychosocial risk enforcement outcomes*. Government of New South Wales.

Harm Alone Is Not Enough 18

Why Some Psychosocial Prosecutions Fail

THE DANGER OF OUTCOME-DRIVEN REASONING

One of the most persistent misunderstandings about psychosocial prosecutions is the belief that harm itself establishes liability. From a human perspective, this belief is understandable. When a worker is seriously psychologically injured, there is a strong instinct to look for fault.

Courts resist that instinct.

This chapter examines a case in which a worker suffered genuine psychological harm, yet the prosecution failed. The court's reasoning is instructive because it clarifies the limits of duty and reinforces that work health and safety law is not a guarantee against all harm.

THE ORGANISATIONAL CONTEXT

The organisation operated in a high-demand environment with fluctuating workloads and exposure to emotionally challenging material. The work was inherently stressful. This was not disputed.

 DOI: 10.1201/9781042031993-21

What distinguished this organisation from those examined in earlier chapters was the way it approached risk management. The organisation had:

- Conducted psychosocial risk assessments
- Implemented workload monitoring processes
- Trained managers in recognising stress indicators
- Provided access to support services
- Adjusted work arrangements where concerns were identified

These measures were not theoretical. Evidence showed they were actively used.

THE HARM

Despite these measures, a worker developed a psychological injury. Medical evidence supported the diagnosis and linked the injury, at least in part, to workplace stressors.

The prosecution argued that the existence of harm demonstrated failure. The defence argued that harm can occur even where reasonable controls are in place.

The court agreed with the defence.

INVESTIGATION AND EVIDENTIARY FOCUS

Investigators examined the organisation's systems over time. They looked for warning signs that were ignored, complaints that were dismissed, or controls that existed only on paper.

What they found was different from earlier cases:

- Concerns raised by the worker were acknowledged
- Adjustments were made where feasible
- Managers sought guidance when issues escalated
- Officers received and reviewed psychosocial risk reports

The evidence showed engagement rather than avoidance.

THE CHARGE AND LEGAL FRAMEWORK

The organisation was charged under the duty provisions of the relevant Work Health and Safety legislation for failing to ensure psychological health so far as was reasonably practicable.

As in other cases, health included psychological health (see *Work Health and Safety Act 2011 (Qld)*, s 4).

The legal question was not whether harm occurred, but whether the organisation failed to do what was reasonably practicable to prevent it.

THE PROSECUTION CASE: CONTROLS WERE INSUFFICIENT

The prosecution argued that:

- Workload pressures remained high
- Stress indicators persisted
- Further controls could have been implemented

The prosecution attempted to frame the organisation's response as reactive rather than preventative.

This argument placed the court in a difficult position: determining how much is enough when risk cannot be eliminated.

DEFENCE SUBMISSIONS

The defence argued that:

- Psychosocial risk cannot be entirely removed in high-demand work
- The organisation identified foreseeable risks
- Reasonable control measures were implemented
- Those measures were reviewed and adjusted over time

The defence emphasised that safety law requires reasonable practicability, not perfection (Safe Work Australia, 2023).

JUDICIAL REASONING

The court's reasoning was measured and precise.

It accepted that the worker suffered genuine harm. It also accepted that work contributed to that harm. What it did not accept was that the organisation failed in its duty.

The court emphasised that:

- The organisation identified the relevant hazards
- Controls were proportionate to the risk
- Responses were timely and documented
- Officers engaged with the issue rather than ignoring it

Crucially, the court held that the presence of harm does not retrospectively render controls inadequate. Reasonableness must be assessed prospectively, not with the benefit of hindsight.

THE OUTCOME

The prosecution failed. The organisation was acquitted.

The decision reinforced that safety law does not impose strict liability for harm. It imposes a duty to manage risk reasonably.

WHY THIS CASE MATTERS

This chapter provides balance to Part Three.

It demonstrates that courts are not predisposed to punish organisations for psychosocial harm where genuine, reasonable efforts have been made. This matters for credibility. A system that punishes regardless of effort would quickly lose legitimacy.

For duty holders, the lesson is constructive. Courts reward engagement, documentation, and responsiveness. They do not require elimination of stress, only its reasonable management.

This case clarifies the boundary of liability: harm triggers scrutiny, not automatic conviction.

The broader message is stabilising. Psychosocial prosecutions are not arbitrary. They track the same principles as physical safety prosecutions. Where genuine, documented, and proportionate effort has been made, courts are reluctant to impose liability even when harm occurs. The goal is not unreachable. It requires genuine engagement rather than symbolic compliance. This case shows what genuine engagement looks like and confirms that it is, in fact, a defence.

For practitioners, this case defines defensible compliance: assessment, proportionate control, monitoring, responsiveness, and contemporaneous documentation of each element. An organisation that can demonstrate all four, consistently and in advance of harm, has met its legal obligation. The bar is high but achievable. It describes the conduct of a genuinely attentive organisation, not a perfect one.

The prospective standard prevents WHS law from becoming outcome-driven. Courts assess reasonable practicability by reference to what was known, and what a reasonable duty holder would have done, at the time decisions were made, not with the benefit of hindsight about what harm followed. This principle protects organisations that genuinely engage with risk from liability for harm they could not reasonably have prevented.

This case is essential to the framework's credibility. Courts do not impose strict liability for psychosocial harm. They impose a duty to manage foreseeable risk reasonably. Where that duty is genuinely met, courts step back even where harm occurs. An acquittal in these circumstances is not a failure of the regulatory system. It is evidence that the system is designed correctly: liability for failure, not liability for outcomes.

REFERENCES

Work Health and Safety Act 2011 (Qld), s 4.

Safe Work Australia (2023) *Managing psychosocial hazards at work—Code of Practice*. Commonwealth of Australia.

Control Without Consistency 19

When Partial Compliance Becomes Evidence of Failure

THE ILLUSION OF SELECTIVE CONTROL

One of the more subtle failure modes in psychosocial risk management is inconsistency. Unlike total absence of controls, inconsistency creates the appearance of effort. Some teams are supported. Some managers respond appropriately. Some complaints are handled well. Others are not.

This chapter examines a psychosocial prosecution in which the organisation pointed to multiple examples of good practice, yet was still convicted. The court's reasoning is instructive: control measures that operate sporadically are not controls at all.

THE ORGANISATIONAL CONTEXT

The organisation operated across multiple departments with significant autonomy at the managerial level. Psychosocial policies existed, including

DOI: 10.1201/9781042031993-22

procedures for managing workload, responding to complaints, and supporting workers experiencing distress.

Training had been delivered. Guidance materials circulated. At a governance level, the organisation appeared attentive to psychosocial risk.

The problem lay in execution.

THE HARM

A worker suffered serious psychological injury after prolonged exposure to unreasonable workload and inconsistent managerial response. In some instances, concerns were acknowledged. In others, they were dismissed or minimised.

The worker experienced a cycle of partial relief followed by renewed pressure. This inconsistency compounded harm rather than alleviating it.

Medical evidence linked the injury not only to workload, but to the unpredictability of organisational response.

INVESTIGATION AND FINDINGS

Investigators examined how psychosocial controls operated across different parts of the organisation. They found:

- Variability in how managers interpreted policy
- Inconsistent escalation of concerns
- Uneven access to adjustments and support
- A lack of central oversight ensuring consistency

Crucially, officers were aware that practices differed between teams. This knowledge transformed inconsistency into a governance issue.

THE CHARGE AND LEGAL FRAMEWORK

The organisation was charged under the duty provisions of the relevant work health and safety legislation (*Occupational Health and Safety Act 2004 (Vic)*, s 21) for failing to ensure psychological health so far as was reasonably practicable.

The prosecution did not allege absence of controls. It alleged failure to ensure controls were applied consistently and effectively.

THE PROSECUTION CASE: VARIABILITY AS RISK

The prosecution argued that:

- Psychosocial risk was foreseeable across the organisation
- Controls were inconsistently applied
- This inconsistency exposed workers to harm
- Officers failed to ensure uniform implementation

The prosecution emphasised that where policies allow discretion, governance systems must ensure that discretion does not undermine safety (WorkSafe Victoria, 2023).

DEFENCE SUBMISSIONS

The defence relied heavily on examples of good practice. It argued that the organisation could not be held responsible for isolated managerial shortcomings where systems existed.

The court rejected this framing. It held that isolated success does not negate systemic failure. Where inconsistency is widespread and known, it becomes a failure of oversight.

JUDICIAL REASONING

The court held that consistency is an essential feature of reasonably practicable control. Controls that depend on individual manager discretion without effective oversight are inherently unreliable.

The court noted that variability itself can be a psychosocial hazard. Uncertainty about how concerns will be treated discourages reporting and exacerbates stress.

THE FINDING AND PENALTY

The organisation was convicted.

The penalty reflected:

- The seriousness of the harm
- The duration of exposure
- The known inconsistency of controls
- The failure to intervene at a governance level

WHY THIS CASE MATTERS

This chapter addresses a blind spot in many organisations.

For duty holders, the lesson is clear. It is not enough to design good systems. Those systems must operate predictably across the organisation. Where outcomes depend on who a worker reports to, risk has not been controlled.

Reasonably practicable psychosocial risk management requires consistency, not pockets of excellence.

This case connects directly to the officer due diligence cases in Part Two. The core failure in Chapter 19 is a governance failure: officers who knew the system was inconsistently applied and did not act. The same obligation that requires officers to verify physical safety systems requires them to verify psychosocial ones. Knowledge of a gap, combined with the capacity and authority to close it, creates the obligation to do so. Inaction is a decision, and courts hold decision-makers accountable for the decisions they make through inaction as much as through action.

The practical challenge for large organisations with decentralised management is ensuring that psychosocial policies operate consistently across diverse teams and managers. Distributing a policy document is the beginning, not the end. Consistent application requires manager training in practical application of the policy, auditing of that application across teams, and meaningful consequences for material deviation. Officers who know that practice varies significantly and take no steps to enforce consistency are failing their verification obligations.

The court's finding that variability itself can be a psychosocial hazard reflects a sophisticated understanding of how organisations cause harm.

When workers cannot predict whether their concerns will be taken seriously, that unpredictability is a stressor. It discourages reporting. It creates an environment in which workers manage risk privately. The resulting absence of safety information is not evidence of safety. It is evidence that the information system is not working.

Inconsistency as a control failure is subtle but legally significant. An organisation with excellent psychosocial controls in some teams and inadequate ones in others has not achieved partial compliance. It has created arbitrary exposure: whether a worker is protected depends on where they sit in the structure, not on the organisation's actual commitment to their safety. Courts treat that arbitrariness as a systemic failure, not a collection of individual team failures.

REFERENCES

WorkSafe Victoria (2023) *Psychosocial hazard prosecution outcome.* Victorian Government.

Occupational Health and Safety Act 2004 (Vic), s 21.

The Point Where Courts Stop 20

Drawing the Boundary of Reasonably Practicable

WHY THE BOUNDARY MATTERS?

This final chapter of Part Three does not examine a single case. Instead, it synthesises the reasoning that courts apply when deciding how far duty extends in psychosocial health (see *Work Health and Safety Act 2011 (Qld)*, s 4).

Understanding this boundary is essential. Without it, leaders either overreact, creating unsustainable systems, or underreact, assuming psychosocial risk is too vague to regulate.

Courts reject both extremes.

WHAT COURTS CONSISTENTLY REQUIRE

Across successful psychosocial prosecutions, courts look for a consistent set of elements:

- Identifiable psychosocial hazards
- Credible indicators of harm or risk
- Knowledge at management or officer level

DOI: 10.1201/9781042031993-23

- Available and reasonable controls
- Failure to implement or enforce those controls

Where these elements align, liability follows.

WHERE COURTS DRAW THE LINE

Equally important are the limits courts impose.

Courts do **not** require:

- Elimination of all stress
- Prevention of all psychological injury
- Prediction of idiosyncratic responses
- Intervention in every interpersonal difficulty

Courts repeatedly emphasise that work health and safety law is not a guarantee of wellbeing. It is a framework for managing foreseeable risk.

REASONABLENESS AS A FORWARD-LOOKING TEST

A critical theme across Part Three is the rejection of hindsight bias. Courts assess reasonably practicable measures based on what was known or should have been known **before** harm occurred.

Controls are assessed prospectively. The fact that harm occurred does not, of itself, prove failure.

This principle underpins acquittals as much as convictions.

THE ROLE OF OFFICERS AND GOVERNANCE

Courts consistently return to governance. Where officers:

- Engage with psychosocial risk
- Seek information beyond reassurance

- Resource control measures
- Respond to emerging signals

Courts are reluctant to impose liability.

Where officers defer, minimise, or fragment responsibility, courts intervene.

THE FINAL JUDICIAL POSITION

Across jurisdictions, courts have settled on a pragmatic position. Psychosocial risk is real, foreseeable, and controllable. It must be managed with the same seriousness as physical risk, but with sensitivity to complexity (Safe Work Australia, 2023).

The law does not demand perfection. It demands effort, responsiveness, and proportionality.

WHY THIS CHAPTER MATTERS

This chapter closes Part Three by clarifying the legal perimeter.

For leaders, the message is stabilising rather than threatening. Psychosocial prosecutions are not arbitrary. They succeed when systems fail, not when harm simply occurs.

Reasonably practicable measures are those that a careful, informed organisation would implement in advance and maintain consistently.

This boundary is where courts stop, and where responsible governance begins.

The final judicial position across all 20 chapters is consistent. Courts expect duty holders to know what risks their work creates, to control those risks as far as reasonably practicable, and to verify that controls are working before harm occurs. The range of hazards covered in this book is wide, from cranes and forklifts to confined spaces, improvised equipment, officer governance, and workplace culture. The underlying legal logic does not vary. Australian courts apply the same scrutiny across all of it. Leaders who understand that scrutiny can engage with safety not as a compliance burden but as a principled governance obligation with knowable, predictable boundaries. That is the real meaning of reasonably practicable.

For practitioners advising on psychosocial compliance, this chapter provides the framework for defensible practice. Before any regulator appears, the organisation should be able to answer four questions: what psychosocial hazards were identified and when? What controls were implemented? How were those controls monitored? How did the organisation respond when a control proved insufficient? If each answer is credible and contemporaneously documented, the organisation has met its legal obligation.

The distinction between acquittal and conviction in psychosocial cases frequently comes down to a single operational question: what did the organisation do when it received information that risk was present? Organisations that assessed the information, adjusted their controls, and documented their response tend to prevail. Organisations that acknowledged the information and continued unchanged tend not to. The difference is action, not intention or goodwill.

The synthesis in this chapter is a clarification of expectations, not a relaxation of them. Courts require genuine engagement with foreseeable risk, proportionate controls, and responsive management when evidence of inadequacy emerges. They do not require perfection or the elimination of all workplace stressors. The standard is reasonable, but it is not low.

REFERENCES

Safe Work Australia (2023) *Managing psychosocial hazards at work—Code of Practice*. Commonwealth of Australia.

Work Health and Safety Act 2011 (Qld), s 4.

Epilogue

WHAT COURTS ARE REALLY SAYING

This book began with fatal incidents involving machinery, gravity, and physical force. It ends with harm that leaves no visible mark. That progression is not accidental. It reflects how Australian courts understand work health and safety: not as a collection of hazards, but as a system of responsibility.

Across all 20 chapters, the same logic appears again and again. Courts are not primarily interested in whether harm occurred. Harm is the trigger for scrutiny, not the proof of failure. What courts examine is decision-making before harm: what was known, what could reasonably have been done, and whether those with authority acted on that knowledge.

In fatality and industrial manslaughter cases, the failures were stark. Hazards were obvious. Controls were basic. Systems relied on luck rather than design. In officer prosecutions, the analysis shifted upward. Courts examined not tasks, but governance: what officers knew, how they assured themselves that systems worked, and whether they were willing to intervene when assurances conflicted with reality. In psychosocial cases, the focus widened again. Courts assessed culture, workload, behaviour, and response, not as abstract concepts, but as operational systems capable of causing harm.

What unites these cases is not moral outrage or regulatory overreach. It is restraint. Courts consistently resist hindsight, emotion, and outcome-driven reasoning. They are careful to draw boundaries. They acquit where duty holders have engaged meaningfully with risk, even when harm occurs. They convict where harm was foreseeable and allowed to persist through inaction, delay, or misplaced confidence.

The phrase 'reasonably practicable' is often misunderstood as vague or elastic. In practice, courts apply it with remarkable consistency. Reasonableness is assessed by reference to knowledge, availability of controls, and proportionality. Practicability is assessed by reference to capacity and opportunity. Where leaders know of risk and have the means to reduce it, courts expect them to act.

This book does not argue that leaders should be punished more often. It shows that leaders are punished predictably. Liability follows patterns. Officers are not prosecuted because they hold titles. They are prosecuted because they control resources, influence systems, and shape priorities. Organisations are not convicted because harm is tragic. They are convicted because risk was normalised, warnings were ignored, or controls existed only on paper.

If there is a single lesson running through every chapter, it is this: safety law does not demand heroism. It demands attentiveness. It demands curiosity. It demands the willingness to look past reassurance and ask whether systems actually work.

Courts are not asking organisations to prevent all harm. They are asking them to take harm seriously before it occurs. Where that happens, courts step back. Where it does not, courts step in.

That is the real meaning of reasonably practicable.

References

Safe Work Australia (2023) Managing psychosocial hazards at work—Code of Practice. Commonwealth of Australia.

Safe Work Australia (2026) Officer duties. https://www.safeworkaustralia.gov.au/law-and-regulation/duties-under-whs-laws/officer-duties

Work Health Authority v Kalidonis NT Pty Ltd (2024) NTLC 4.

Kalidonis NT Pty Ltd v Work Health Authority (2025) NTSC 28.

Work Health and Safety (National Uniform Legislation) Act 2011 (NT), ss 19, 32.

SafeWork SA (2022) National Petroleum & Corrosion Services Pty Ltd—Prosecution outcome. Government of South Australia.

Office of the Work Health and Safety Prosecutor (2024) *R v Narellan Pools Pty Ltd—Industrial manslaughter conviction*. Queensland Government.

Office of the Work Health and Safety Prosecutor (2022) *R v Owen—Industrial manslaughter prosecution and sentencing outcome*. Queensland Government.

Office of the Work Health and Safety Prosecutor (2024) *Industrial manslaughter and fatality prosecution summaries*. Queensland Government.

WorkSafe WA (2025) RPC Surface Treatment Pty Ltd—Prosecution summary and sentencing outcome. Government of Western Australia.

Safe Work Australia (2022) *Psychological health and safety at work*. Commonwealth of Australia.

Safe Work Australia (2023) *Interpretive guideline: The meaning of "reasonably practicable"*. Commonwealth of Australia.

SafeWork NSW (2022) *Officer due diligence prosecution outcomes*. Government of New South Wales.

SafeWork NSW (2023) *Psychosocial risk enforcement outcomes*. Government of New South Wales.

SafeWork SA (2024) Service pit death brings $300,000 fine. Government of South Australia.

SafeWork SA (2024) Apprentice death leads to record work safety fine. Government of South Australia.

District Court of Queensland (2020) *R v Brisbane Auto Recycling Pty Ltd* [2020] QDC 113.

Occupational Health and Safety Act 2004 (Vic), s 21.

Work Health and Safety Act 2011 (Qld).

Work Health and Safety Act 2012 (SA), ss 19, 32.

Work Health and Safety Regulations 2012 (SA), Part 4.3.

Work Health and Safety Act 2020 (WA).

WorkSafe Victoria (2022) *LH Holding Management Pty Ltd—Prosecution outcome*. Victorian Government.

WorkSafe Victoria (2023) *Psychosocial hazard prosecution outcomes.* Victorian Government.

WorkSafe WA (2022) *M Construction (WA) Pty Ltd—Prosecution summary.* Government of Western Australia.

Index